The Invention of Modern Science

Edited by

Sandra Buckley

Michael Hardt

Brian Massumi

THEORY OUT OF BOUNDS

19   The Invention of Modern Science   Isabelle Stengers

18   Methodology of the Oppressed   Chela Sandoval

17   Proust and Signs: The Complete Text   Gilles Deleuze

16   Deleuze: The Clamor of Being   Alain Badiou

15   Insurgencies: Constituent Power and the Modern State   Antonio Negri

14   When Pain Strikes   Bill Burns, Cathy Busby, and Kim Sawchuk, editors

13   Critical Environments: Postmodern Theory and the Pragmatics of the "Outside"   Cary Wolfe

12   Metamorphoses of the Body   José Gil

11   The New Spinoza   Warren Montag and Ted Stolze, editors

10   Power and Invention: Situating Science   Isabelle Stengers

9   Arrow of Chaos: Romanticism and Postmodernity   Ira Livingston

8   Becoming-Woman   Camilla Griggers

7   A Potential Politics: Radical Thought in Italy   Paolo Virno and Michael Hardt, editors

6   Capital Times: Tales from the Conquest of Time   Éric Alliez

5   The Year of Passages   Réda Bensmaïa

4   Labor of Dionysus: A Critique of the State-Form   Michael Hardt and Antonio Negri

3   Bad Aboriginal Art: Tradition, Media, and Technological Horizons   Eric Michaels

2   The Cinematic Body   Steven Shaviro

1   The Coming Community   Giorgio Agamben

The Invention of

Modern Science

Isabelle Stengers

Translated by Daniel W. Smith

Theory out of Bounds   *Volume 19*

University of Minnesota Press

Minneapolis • London

This translation would not have been possible
without the support of a Vice-Chancellor's Postdoctoral Fellowship
in the School of Philosophy at the University of New South Wales,
whose generosity is gratefully acknowledged.

Originally published as *L'Invention des sciences modernes* (Paris: La Découverte, 1993).
Copyright 1993 Gius. Laterza & Figli Spa, Roma-Bari. This book is a result of
the cooperation between Laterza Publishing House and Sigma Tau Foundation
in the "Lezioni Italiane" series. The English-language edition has been
arranged through the mediation of Eulama Literary Agency.

Published by the University of Minnesota Press
111 Third Avenue South, Suite 290
Minneapolis, MN 55401-2520
http://www.upress.umn.edu

Printed in the United States of America on acid-free paper

LIBRARY OF CONGRESS CATALOGING-IN-PUBLICATION DATA
Stengers, Isabelle.
[Invention des sciences modernes. English]
The invention of modern science / Isabelle Stengers ; translated by
Daniel W. Smith.
p.   cm. — (Theory out of bounds ; v. 19)
Includes bibliographical references (p.   ) and index.
ISBN 0-8166-3055-0 (hc : acid-free paper) — ISBN 0-8166-3056-9 (pb :
acid-free paper)
1. Discoveries in science.   2. Science—Philosophy.   I. Title.
II. Series.
Q180.55.D57 S7413 2000
501—dc21
00-008711

The University of Minnesota
is an equal-opportunity educator and employer.

11 10 09 08 07 06 05 04 03 02 01 00   10 9 8 7 6 5 4 3 2 1

For Félix Guattari and Bruno Latour,
in memory of an encounter that never took place.

Contents

PART I.   **Explorations**   1

Chapter 1.   **The Sciences and Their Interpreters**   3
Scandals—Autonomy—A Destructive Science?—The Leibnizian Constraint

Chapter 2.   **Science and Nonscience**   21
In the Name of Science—Break or Demarcation?—Popper's Question—
The Unfindable Criterion—One Historical Tradition among Others?

Chapter 3.   **The Force of History**   39
The Singularity of the History of the Sciences—The Three Worlds—
Clarifying the Paradigm

PART II.   **Construction**   55

Chapter 4.   **Irony and Humor**   57
Constructing a Difference—Great Divisions—The Political Invention of
the Sciences—On the Event

Chapter 5.   **Science under the Sign of the Event**   71
In Search of a Recommencement—The Power of Fiction—A New Use of
Reason?—The Inclined Plane

Chapter 6.   **Making History**   89
Negative Truth—Authors to Interest—Bringing into Existence—Mediators—
Political Questions

PART III.   **Propositions**   109

Chapter 7.   **An Available World?**   111
The Power in Histories—Mobilization—The Patron's Job—The Politics
of Networks

Chapter 8.   **Subject and Object**   131
         What Singularity for the Sciences?—Mathematical Fictions—Darwin's Heirs—
         Demoralizing History—"What Does He Want from Me?"

Chapter 9.   **Becomings**   151
         How to Resist?—Nomads of the Third World—The Production of Expertise—
         A Return to the Sophists

      Notes   169

      Index   179

Explorations

O   N   E

The Sciences and Their Interpreters

Scandals

A DISTURBING rumor has been spreading in the world of scientists. It seems that there are some researchers—specialists in the human sciences, no less—who are challenging the ideal of a pure science. A field is being constituted, born in England some twenty years ago,[1] prospering in the Anglo-Saxon countries, but now present in France.[2] This field, known under various names—"social studies in science," "sociology of the sciences," "anthropology of the sciences"—puts in question any separation between the sciences and society. The researchers it brings together would dare to claim to study science as a social undertaking like any other, neither more detached from the cares of the world nor more universal or rational than any other practice. They no longer denounce the numerous infidelities committed by scientists against their own norms of autonomy and objectivity, but consider these very norms to be empty, as if every science were "impure" by nature and not because of its deviation from an ideal.

The thinkers of science sharpen their weapons and rise to the defense of a threatened cause. Some of them rely on the very classical argument of retaliation [rétorsion]. It has been useful, though it still keeps to old paths. In saying that science is a social undertaking, doesn't one subordinate it to the categories of sociology? Now, sociology is a science, and in this case it is a science that is trying to become a superscience, the science that explains all others. But how could it escape the very disqualification it brings on the other sciences? Thus, sociology disqualifies itself and cannot claim to impose its interpretive grid. Others play the card of realism: if everything is only a social bond, that is, conventional and arbitrary, how have we been able to send men to the moon (and, one might add, to explode atom bombs)? When the need arises, don't sociologists of science, like everyone else, run to the doctor, who prescribes vaccines and antibiotics, which are products of science? Others suggest that the putting in question of scientific objectivity be assimilated to the justification of a brutal law of the strongest. Civilization is in danger!

The anxiety of the scientific world is strange, because it repeats, as if in a delayed reaction, the anxiety that seized the small world of philosophers of science when the historian Thomas Kuhn proposed, in 1962, the category of "normal science." No, asserted Kuhn, the practicing scientist of a given science is not the glorious illustration of a critical mind and the lucid rationality, which these philosophers tried to characterize through the scientist. Scientists do what they have learned to do. They treat the phenomena that seem to be the concern of their discipline in accordance with a "paradigm"—which is both a practical and a theoretical model—that seems obvious to them, and in relation to which they have very little distance. Worse, since every paradigm determines the legitimate questions and the criteria according to which responses can be recognized as acceptable, it is impossible to construct a third position "outside the paradigm" from whence the philosopher would be able to evaluate the respective merits of rival interpretations (thesis of noncommensurability). Worse yet, the scientist's subordination to the paradigm of his community is not a fault. According to Kuhn, what we call "scientific progress"—the cumulative process thanks to which ever more phenomena are becoming intelligible, technologically controllable, and theoretically interpretable—depends on it. And he describes in harsh terms the lucidity of scientists who belong to disciplines without a paradigm: either they argue with each other, tear each other apart, accuse one another of ideological biases, or else they coexist in the indifference of schools sanctioned by the names of their founders. We speak of "Piagetian" psychology, "Saussurean" linguistics, "Lévi-Straussian" ethnology, and the very adjective signals to their happy colleagues that here science does not have the power to make scientists agree. We do not speak of "Crickian" biology or "Heisenbergian" quantum mechanics, do we?

The philosophers of science exhibited a considerable discontent. Of course, they reverted to the argument of retaliation [*rétorsion*]: Kuhn proposes the idea of a "paradigm" to the historian and philosopher of science, and thus he has no right, according to these very terms, to claim to describe the sciences "as they really are." They claimed that it was impossible to put an outmoded science, like the one that thought of water as an element, on the same level as today's science, which water confirms by letting itself be ruthlessly synthesized and decomposed. They denounced the tragedy for civilization entailed by reduction of science to *mob psychology*, to a psychology of irrational crowds subject to the effects of fashion and imitation.

The majority of scientists, however, did not have the same reaction at all. They *loved* Kuhn's "paradigms" enormously. They recognized in them a pertinent description of their activity. The notion of a "paradigmatic revolution," in which one paradigm displaces another, was for them an appropriate way to describe their discipline's history. And the human sciences began to dream of a paradigm that would one day bestow on them the progressive mode of their happy colleagues. "New paradigms" began to flourish almost everywhere, from systems theory to anthropology or sociology.

Why did something that scandalized the philosophers satisfy so many scientists? And why are they so scandalized now? Hadn't Kuhn already stressed the social dimension of the sciences by showing that the scientist must be described as a member of a community and not as a rational and lucid individual? It is the question of this curious delayed reaction that will be my point of departure.

Autonomy

We can, I believe, affirm that, from the viewpoint of scientists, Kuhn's description preserves the essential thing: the autonomy of a scientific community in relation to its political and social environment. This autonomy does more than simply preserve the community; it institutes this community as the norm and as the condition of possibility for the fruitful exercise of a science, whether it is a question of the practice of a normal science or the paradigmatic revolutions that rejuvenate it. Not only are scientists not asked to give an account of their choices and research priorities, but it is just and normal that they are unable to provide such an account. What makes the paradigm so fruitful is its largely tacit character, transmitted by the pedagogical artifice of textbook examples and problems to resolve. Because they do not need to effect a critical distancing, scientists can confront the most disconcerting phenomena with confidence, deciphering them without vertigo, in the mode of resemblance to their paradigmatic object. Moreover, this confidence also explains the fruitful scandal that Kuhn associated with the notion of the "anomaly," a stumbling point where a difference is recognized as significant, putting in question not the scientist's competence, but the paradigm itself.

According to Kuhn, the paradigm thus explains not only the cumulative conquest, but also the invention of the new. The anomaly, both an agent provocateur and a point of fixation, puts the scientist in a state of "tension"; he or she becomes the vector of a creativity that perhaps would not have inspired a lucid—that is, skeptical—attitude toward the power of theories. Correlatively, the indifference of the community toward difficulties or incomprehensible results is justified. In itself, no brute, abnormal "fact" has the power to be recognized as an anomaly. And no anomaly gives the person who recognizes it the power to claim the attention of the collectivity. The "paradigmatic crisis" becomes collective when the scientist has gained the power to counterinterpret the results of his colleagues, when a new paradigm, the bearer of a new type of intelligibility, imposes a choice. Lucidity is the result of a crisis; it must be conquered and cannot be considered normal.

The reading proposed by Thomas Kuhn thus justifies a radical differentiation between a scientific community, produced by its own history, endowed with instruments that inseparably integrate production (research) and reproduction (the training of those authorized to participate in this research), and a milieu that, if it wishes to benefit from the repercussions of this activity, must be content with maintaining it without making it give an account of itself. With regard to the scientist at work, no one has to benefit from a relation of force that would allow him or her to impose questions that are not the "good" questions of the community. Every attack on the autonomy of a community working under the paradigm amounts to "killing the goose with the golden eggs," to attacking the very condition of possibility of scientific progress.

In fact, Thomas Kuhn did not invent this argument, which prohibits one from asking scientists to give an account of their choices and priorities. In 1958, the physicist Michael Polanyi had already linked the fruitfulness of scientific research to a "tacit knowledge," which is very different from knowledge that deals with the explicit or explainable contents of science. Polanyi's scientist is similar to a connoisseur, in the English sense of an "expert," and his competence is inseparable from an engagement that implies intelligence, but also gestures, perception, passion, belief.[3]

Polanyi emphasized the "phenomenological" description of the scientist at work more than the way scientific communities ensure the transmission of their mode of engagement. But for all that, his position was not devoid of any sociopolitical preoccupation. On the contrary: his work was inscribed in a debate that was set off in England on the occasion of the Second International Congress on the History of Science and Technology (London, 1931). During this congress, Nikolay Bukharin, the head of the Russian delegation, had pointed to the "absolutely new perspectives" opened up in his country by the rational implementation of scientific production within the framework of a planned economy.[4] Young Marxist scientists such as John D. Bernal and Joseph Needham had been enthusiastic about this perspective. In 1939, Bernal published his *The Social Function of Science*, which presents scientific production and social and economic interests as being bound up with each other in fact and in principle.[5] Bernal concluded that a profound reorganization of science was necessary, one that would make it able to respond to true social needs. It was against this "Bernalism" that Michael Polanyi, at the beginning of the war, created the Society for Freedom in Science.

After the war, the debate was again taken up, but this time the threat no longer came from Marxist intellectuals. It was a matter of protesting against the projects of planning scientific choices by Western governments. In 1962, Polanyi published a doctrinal article, "The Republic of Science," which explicitly linked the claim of science's "extraterritoriality" with the figure of the "competent" scientist, who alone is capable of evaluating research in his own domain, without for all that being able to give an account of his evaluative criteria.[6] More precisely, Polanyi held that scientific communities realize a principle that "in its highest sense" is *reduced* to the market mechanism when applied to economic activities. Every scientist is inserted into a network of mutual appreciations that extend well beyond his or her own horizon of competence. The Republic of Science

> must advance by supporting independent initiatives, coordinating themselves mutually to each other. Such adjustment may include rivalries and opposing responses which, in society as a whole, will be far more frequent than they are within science. Even so, all these independent initiatives must accept for their guidance a traditional authority, enforcing its own self-renewal by cultivating originality among its followers.[7]

I will not here recount this entire history, which involves, on the one hand, the question of the Marxist and then Stalinist conceptions of science (or the theses on bourgeois science and proletarian science in postwar France), and, on the other hand, the historians' debate concerning the "internal" and "external" history of the sciences, with which names such as Alexandre Koyré and Charles Gillispie are associated. I will content myself with emphasizing that the defense of the "internal" history—which holds that scientific knowledge develops according to its own criteria, "external" factors merely playing a subordinate role—must not be confused with the defense of a "rational" science, in the sense in which most philosophers of science understood it at the time. This is what Polanyi's "postcritical" philosophy asserted. And this is what Kuhn's *The Structure of Scientific Revolutions* would make more explicit.

The novelty of Thomas Kuhn's work is thus completely relative. Above all, it lies in his explanation of the divergence between the interests of scientists and those of philosophers of science. The former have no need to pass through the defense and illustration of the rationality of the sciences in order to claim the initiative of posing questions and exclusivity in judgments of value and priority. The latter thus lose any privileged status: they are neither arbiters nor witnesses, nor are they even able to decipher the norms that function implicitly in the sciences and that allow science to be distinguished from nonscience.

What, then, is the new "anthropology" or "social history" that so scandalizes scientists? It is inscribed explicitly in the wake of Kuhn's work, but it does not display the same respect for scientific productivity. A new discourse is being constructed, which explicitly distinguishes between the things that interest scientists and the things that interest the people who study scientists. The latter, if they want to be recognized as legitimate participants in the new field, must be subjected to a discipline named "the principle of symmetry." It is a matter of drawing conclusions from the fact that no general methodological norm can justify the difference between winners and losers, which creates the closure of a controversy. Kuhn here relied on a certain rationality of scientists, who evaluate the fruitfulness and power of the competing paradigms. The difference, for him, was by no means arbitrary. The principle of symmetry requires that one no longer rely on the hypothesis of this rationality, which leads the historian to borrow the vocabulary of the winner when recounting the history of a controversy. On the contrary, what must be brought to light is the situation of fundamental indecision, that is, the set of possibly "non-scientific" factors that were at play in the creation of the final relation of force, which we inherit when we think that the crisis has created the difference between winners and losers.

The paradigm guaranteed the communities' autonomy and was content to interpret what traditionally characterized the ideal of a "true" science in a different manner: cumulative progress, the possibility of consensus, the irreversibility of the distinction between the outdated past and the unknown future. The principle of symmetry requires the researcher to be attentive to everything that, again in traditional terms, was judged to be a gap or a lack in relation to this ideal: relations of force and overtly social games of power, differences in resources and prestige between competing laboratories, possibilities of being aligned with "impure" interests (ideological, industrial, state, etc.). Whereas the image of the sciences constructed by Polanyi corresponded to the ideal free market, the Kuhnian image of science, which is less focused on the individual scientist, is closer to the Hegelian idea of "the cunning of reason": through "irrational" means, a history is constructed that corresponds, point by point, in an optimal manner, with what we would expect from an undertaking with a rational motor. The new image associated with the sociology of the sciences brings to light our inability to judge the history we have inherited. Because we are the heirs of the winners, we re-create, with regard to the past, a narrative in which arguments internal to a scientific community would be sufficient to designate these winners; it is because these arguments convince us as heirs that we retrospectively attribute to them the power to have made the difference.

Correlatively, there is the theme of the "great division," the difference between the "four European centuries," during which time modern science was created, and all other civilizations, which lose the event-like character conferred on them by Kuhn and the group of "internalist" historians. According to Kuhn, it was here, and nowhere else, that the condition of possibility for science was realized—namely, in the existence of societies that give scientific communities the means of existing and working without intervening in their debates. But these four centuries have been marked by other singular innovations. Do not industry, the state, the army, and commerce all enter into the history of scientific communities on two fronts, both as sources of financing and as beneficiaries of the useful results? Questions about the "external" history of the sciences reappear here, but in a much more formidable manner. It is no longer a question of a general thesis concerning the interdependence of scientific practices and their environment. Like all other human beings, scientists are the product of a social, technological, economic, and political history. They actively draw on the resources of this environment to win acceptance for their theses, and they *conceal* their strategies under the mask of objectivity. In other words, as the product of their epoch, scientists have become actors; and if, as Einstein said, it is not necessary to rely on what they say they do, but to look at what they do, this is in no way because scientific invention goes beyond words, but because words have a strategic function that one must know how to decipher. Scientists, rather than heroically depriving themselves of any recourse to political authority or the public, here seem to be accompanied by a cohort of allies—anyone whose interest was able to make a difference in the controversies that set them against their rivals.

A Destructive Science?

Most "relativist" sociologists deny any will to "denounce" science. They merely want to do their job, which presupposes a difference in principle between the interpretation a social practice gives of itself and the one constructed by a sociologist. In principle, scientists should be no more scandalized than any other social or professional group that becomes an object of interest for sociologists; and if they are, they give themselves away, they admit to claiming an undue authority for themselves, thereby confirming the legitimacy of the inquiry. Yet it is here that the argument of retaliation [*rétorsion*] can be applied: Is not sociology itself a science? By what right, if not in the name of science, can sociologists ignore the fact that, of all the interpretations given to science, it is their own that clash most painfully with the scientists'? For, to be sure, the sociologist is not alone in interpreting scientific practices; there are others that question the meaning and stakes of the sciences in a more determined manner. I will take as my examples the critique of science as a "technoscience" and the radical feminist critique of scientific rationality, and I will attempt a first characterization of the sciences on the basis of this initial problem: Why are some interpretations that question scientific rationality far less disturbing to scientists than others?

One might think that scientists would protest unanimously against the staging of the relation of radical opposition between "science" and "human culture," on which the critique of technosciences is based. How can one accept seeing the sciences as the expression of an unbridled rationality, escaping the control of humans, intent on denying, subjecting, and destroying everything it cannot reduce to the calculable and the manipulable? Now, the protests of scientists are rather rare, as if they recognized the painful legitimacy of a hypothesis that celebrates the divorce between their undertaking and Enlightenment values, between the service of science and that of humanity.

The critique of the "technosciences" identifies "scientific rationality" with a purely operative rationality, reducing everything it conquers to a calculus and technical domination. It denies any possibility of distinguishing between scientific, technical, and technological productions, and refers as often to the sociotechnical apparatuses that effectively transform human practices, such as computer science, as to the "scientific visions of the world" that reduce reality, for example, to an exchange of information.

The radical feminist critique begins with the same type of description, but it identifies this rationality not with the destruction of all value, but with the triumph of "male" values. For a long time now, a good number of feminist authors have emphasized the degree to which scientific research is dominated by the ideals of competition, polemical rivalry, sacrificial commitment to an abstract cause, in short, to a form of organization that I will later place under the sign of *mobilization*. However, they did not question the very mode of knowledge invented by the sciences. Rather, they took aim at fields such as medicine, history, biology, or psychology, which are concerned with sexed beings, where it is possible to show that questions can be effectively "biased" by conscious or unconscious presuppositions with regard to women. This critique, which is sometimes termed "empiricist," is opposed to the radical feminist viewpoint, for which the whole of the sciences is a "social-sexed product," the result of a society dominated by men.[8] In this case, from mathematics to chemistry, from physics to molecular biology, nothing must escape the feminist critique.

Both critiques, technoscientific and feminist, assume a perspective of resistance, but in both cases, what they are resisting has been depicted in such a way that the appeal to resistance takes on a prophetic accent. Whether that rationality is "all-encompassing" and endowed with its own dynamic, or translates a sexed mode of relation to the world and others, it has the power to define its actors, and can only be limited, regulated, or transformed from the outside, by a "totally other" free from any compromise. Would an "other" science, feminine or feminist, be possible? The burden of proof falls on women. Scientists, whether they are mocking or sincere, can declare themselves to be extremely interested in the perspective of a different mathematics or physics. Can a new ethical conscience pose a counterweight to technoscientific power? The burden of proof falls on society or the authorities that represent its values, and scientists will not balk at participating in "ethics committees" where they will represent the "ends of science" before various representatives while being confronted with the "ends of humanity."

In fact, the price paid by the radical character of the critique, whether technoscientific or feminist, is the respect it accords the scientist as a privileged interpreter of what science is able to do. Scientific rationality, as here criticized, is not identified with respect for a norm, which could be verified. Rather, it is concerned with a destiny, and it is the truth of this destiny that turns every vision of reality into something manipulable, whatever distance there might be between the claims of this vision and the practices it authorizes. In this sense, the "radical" critique of science grants scientists all their pretensions. It recognizes the sociotechnical mutations that affect our world as the products of rationality—(techno)scientific or male—and tends to accept what scientists "say" at face value, even in their most daring extrapolations. The latter are thus not treated with suspicion, but as truthful witnesses.

We should therefore not be surprised that the question of techno-science, if it arises, can be taken up by scientists. For it casts them in the painful but honorable role of representatives of a radically new mutation, without equivalent in human history, expressions of an inhuman imperative, perhaps, but one that purifies them and preserves them from any vulgar questioning. If technoscience celebrates the terrible dynamic that makes the rational communicate with the irrational, the imperative to control and calculate with the establishment of an autonomous system, uncontrollable from the inside, which makes power and the absence of meaning co-incide, then scientists, technicians, and experts are not subject to questioning, be-cause, like everyone else, they are waiting for limits to the power of expansion of a dynamic that defines them beyond their intentions and their myths.

Correlatively, contrary to the relativist sociologists, the radical critique of the sciences is hardly preoccupied with following the details of scientific controversies or using the "principle of symmetry" between winners and losers. What-ever theses they confront, the moment they fall under "technoscience" (or "male" science), knowing who will win (and how) matters little. In any case, the victory will only sanction a new advance for a purely operative and dominating rationality, which makes truth coincide with the sole criterion of "it works," to the detriment of culture, its values, its significations. This has very concrete consequences for those who, today, in the name of progress or rationality, insist on the necessity of this or that program of research. In particular, they are not concerned, when sitting on "bioethics" com-mittees, for example, with disreputable antiestablishmentarians, who are persuaded a priori that scientists' arguments are in fact relative to the scientists' interests, but with the protagonists who accept, in principle, their status as representatives of an "operative logic," and argue about the possible bounds to set to this logic.

The great difference between the relativist description of scien-tific practices and the radical critique of science thus comes down to a contrast that can be taken as a first approach to the singularity of the sciences. The argument ac-cording to which scientific progress serves the ends of humanity can be used by sci-entists if the case arises, but this argument does not seem to convey the intrinsic meaning they give to their activity. The argument according to which science is a critical and lucid activity is used in certain circumstances, when it is a matter of showing how different it is from astrology or parapsychology, for example, but it can also be abandoned in favor of the representation of a fruitful somnambulist. In contrast, the argument according to which the knowledges produced by the sciences are not relative to situations of social relations of force, and can take advantage of a privileged relation with regard to the phenomena they are dealing with, seems to be crucial. If this relation is not neutral, if it is reducible to the calculable and the controllable, so be it. But to say that it is arbitrary, that it is the simple product of an "understanding" [*entente*] between scientists and demonstrates nothing more than a human convention—this is what is intolerable. If the sciences are full of impurities, and embedded in situations where the effects of fashion and social or economic interests have played a role, so be it. But to deny any distinction between "true science," ideally autonomous in relation to "nonscientific" interests, and the gaps [*écarts*], foreseeable and regrettable, in relation to this ideal—this is what arouses the most scandalized protests.

The specific problem of the relativist sociological approach to the sciences is thus that it seems to have to confront head-on the conception that scientists themselves harbor of science. Certainly this could be a claim to glory. Whereas the radical critique of scientific rationality can, if the case arises, stabilize what it is aiming at in the conviction—or myth—of its formidable yet honorable destiny, here we would finally have the instruments of a veritable contestation of the power of the sciences. But are we so sure of the pertinence of these instruments? Do we really want scientists to be willing to bring together strategies that are indifferent to the "truth," and to be solely interested in allying themselves with the powers that can help them make the difference [*faire la différence*]? In return, do we truly want these powers to be able to require scientists to stop splitting hairs and to align themselves with the demands of normalization, interest, or profitability?[9] In whose name can the claim to autonomy be ridiculed?

We can comprehend, as if it were a "cry," the protest of scientists against the sociologists' approach, as if it were at once the expression of a wound, a revolt, and a disquietude.

A wound, because they well "know" that their activity is only a social activity "like the others," that it is exposed to risks, demands, and passions without which it would only be a bureaucracy of numbers or an obsessive construction of metrological networks. They would be the first to recognize that it is "also" all that, but they know that it is not "only" that.

A revolt, because they are betrayed by those who have at their disposal infinitely more "words," references, and argumentative capacities—that's their job—to take the sciences to task. As long as these "gossips" were using their resources to construct a privileged image of science, the situation was balanced. Scientists could even criticize the all-too-rational image given to their particular science — as Einstein did not hesitate to do. But if, as happens today, those people whose job it is to speak of science turn their argumentative resources "against" the scientists, they are taking advantage, in a revolting manner, of the powers of rhetoric against the reality — mute and probative — of science.

Disquieting, finally, because the rhetorical resources of the discourse on science are one of the resources of science, as much with regard to its internal controversies as to the negotiations between disciplines and frontiers. Recent paradigms, but also, for more than a century, the epistemological distinction between "pure" and "applied" sciences, are among the arguments that allow one to resist, to plea, to protect oneself, to attract interest, to ask for help. If these arguments are deciphered as a strategic resource and not as an epistemologically grounded expression of scientific reality, they will no doubt become unusable. If scientific knowledge is henceforth reputed to no longer be any more disinterested than other knowledges, if it is valid only through the allies it is capable of recruiting, how can minority scientists plead their case?

There is thus a great difference between the respective positions of philosophers and scientists I outlined at the beginning of this chapter. The philosophers were requiring the sciences, which they do not practice, to be such that they justify the practice of the philosopher of the sciences. They were demanding that the sciences illustrate or imply a definition of scientific rationality, which it would be the philosophers' task to disengage, and which would give them the power to know, better than the scientists themselves, what defines scientists as such. One of the risks of the philosopher's job is to be disappointed by what one was hoping to be able to assign the role of the ground [*fondement*]. After the protests and indignations can come the time for the invention of new questions, perhaps more pertinent, perhaps capable of transforming, for better or worse, the disappointment into a problem.

Scientists, by contrast, do not have this liberty. It is they who are being described, it is their activity one is attempting to characterize, and ever since the modern sciences have been imposed as a reference in the landscape of our practices and our knowledges, they have never ceased to be described and characterized in this manner. Most of the time, certainly, description and characterization have been strategic resources for them, but that is not enough to justify, like the return of a well-deserved stick, a description that scandalizes them, that seems to them to deny the truth of their engagement and their passion. And the good intentions of those who intend to "demythify" science are not enough either. Can they guarantee that other protagonists will not be interested in taking them literally, that is, in using their theses in order to put science a little more in the service of their own interests?

The Leibnizian Constraint

No statement, if it is held in the name of truth, or good sense, or the will not to let oneself count on it, is able to control the consequences of its enunciation. This, in any case, is the principle to which I wanted to subject my interpretation of the sciences. More precisely, the latter would have to respond to the "Leibnizian constraint" according to which philosophy should not have as its ideal the "reversal of established sentiments."[10]

Few philosophical statements have been as badly viewed as this one. Even Gilles Deleuze has spoken in this regard of Leibniz's "shameful declaration." And yet it is easy to "speak the truth" against established sentiments, and then to be proud of the effects of hatred, ressentiment, and panicked rigidity one has aroused as so many proofs that one has "reached the beast"—even at the price of persecution, since the martyr and the truth are good bedfellows. Leibniz, the diplomat who desperately sought to create conditions for peace between religions, knew this well, living in a Europe bending under the legacy of so many martyrs. If his aim was to "respect" established sentiments, it seems to me, it was much as a mathematician "respects" the constraints that give meaning and interest to his problem. And this constraint—not to clash with, not to reverse established sentiments—does not mean not to clash with anyone, to make everyone agree. How could Leibniz not have known that the way he used references to the Western tradition was going to clash with all those who made use of "established sentiments" to maintain and stabilize hateful mobilizations? The problem designated by the Leibnizian constraint ties together truth and becoming, and assigns to the statement of what one believes to be true the responsibility not to hinder becoming: not to collide with established sentiments, so as to try to open them to what their established identity led them to refuse, combat, misunderstand.

We should not be overly hasty in identifying this project with a naive optimism. It is more a question of a technical optimism, expressing the technical know-how of the diplomat with regard to the crimes entailed by the heroism of truth. If nature makes no leaps, nothing, as Samuel Butler noted, is more fearsome than humans who believe they have made one, converts who, ferociously or devotedly, turn against those who remain in the illusion from which they have just extracted themselves.[11]

Today, we no longer kill or die to defend scientific objectivity or the right to put it on trial. But the words we use have the power to clash with others, to scandalize, to provoke hateful misunderstandings. In this book, I will dare to associate scientific reason and politics. I know I run the risk of offending those for whom

nothing is more important existentially, intellectually, and *politically* than to maintain a difference between the two. But in the name of this established and eminently
respectable sentiment, must we conserve the categories that, each and every day, give
proof of their own vulnerability? "In the name of science," "in the name of scientific objectivity," we see definitions and redefinitions of problems being constructed
that implicate human history. Is it not necessary to invent words that would permit
this reference to be *rendered discussable*, to make science political?

      The challenge of this book is thus to try to articulate what we
understand by science and what we understand by politics, without clashing with,
not all "sentiments," but what I will call, following Leibniz, the established sentiments,
those that provide a point of reference, that cannot be threatened without leading
to panicked rigidity, indignation, or misunderstanding. To do this, I will try to put
to work what I will call, following Bruno Latour, to whom this book is dedicated, a
"principle of irreduction." This principle constitutes both a "putting on guard" and
and a demand whose target is the set of theses that lend themselves to a slight modification, and indeed, that implicitly call for one: the passage from "this is that" to
"this is not that" or "is only that." To speak of science in a political register, for example, would become "science is only politics," an enterprise in which power is at
stake, protected by an illusory ideology, managing to impose its particular beliefs as
universal truths. On the contrary, to protest that science transcends political divisions
would be to implicitly identify the political register with the arbitrary, tumultuous,
and irrational waves of human controversies that lick the feet of the scientific fortress,
and in some cases, that take elements born in innocence and put them to perverse,
harmful, or irresponsible uses. Each of these theses either asserts a reducibility or
denies the possibility of a reduction in the name of a transcendence, which implies
that the person who is speaking knows what he is talking about, in other words, that
he is himself in the position of a judge. He knows, in this case, what "science" and
"politics" are, and gives or refuses to one of these terms the power to explain the
other. The principle of irreduction prescribes a retreat from this claim to know and
to judge. For what if what we today call "politics" was marked by the tendency to
exclude science from itself, so that what we call the "sciences" had to present themselves as "apolitical"? What are these "words"—objectivity, reality, rationality, truth,
progress—if they are not taken as shams dissimulating one human enterprise "like
any another," nor as guarantees of an essential difference?

      Irreduction thus signifies a certain distrust of all the "words"
that lead quasi-automatically to the temptation to explain by reducing, or to construct
a difference between two terms that reduces them to a relation of irreducible oppo-

sition. In other words — and here again I am appealing to the demand posed by Latour in *We Have Never Been Modern* — it is a matter of learning to use words that do not bestow, as if it were their vocation, the power to *unveil* (the truth behind appearances) or to *denounce* (the appearances that veil the truth). We must be clear that this does not mean we will reach a world where everyone would be beautiful and kind. I hope to make myself hated, but I would like to try not to be hated by those whom I have no desire to offend — that is, all those who *submit to* the mobilizing power of words that recruits them into antagonistic camps, without for all that having an active stake in the maintenance of this antagonism.

  The stakes in an approach to science that respects the "Leibnizian constraint" can also be stated in terms of the mode of laughter that should be "relearned" with regard to science. There was a time, not so long ago, when science was discussed in the salons. During this time, Denis Diderot imagined the mathematician d'Alembert transported by a dream in which he experienced himself as matter, and Dr. Bordeu speaking to Mlle de Lespinasse of "varied and regular attempts" to create, possibly, a race of "foot-goats," intelligent, indefatigable, and fast... who would make excellent servants.[12] What philosopher would today dare the fiction of a mathematician known to be transported by a delirious dream, and who would dare to laugh at the things discussed and regulated by jurists, moralists, theologians, and doctors on what are called "ethics committees"? And yet I have no desire to be mobilized in a denunciative cohort before having learned to laugh, before having learned how not to let myself be redefined as the member of a group with a minority vocation, which itself also seeks to impose its "values," its "imperatives," its "vision of the world." I do not want to have a seat on an "ethical commission" next to a theologian, a psychoanalyst, a philosopher specializing in technoscience, and a doctor who is a mandarin scientist and moralist. I want to be able — and to incite people other than myself to be able — to intervene in this history without arousing a past in which other moral majorities were dominant.

  The king is not naked: more or less everywhere, there are procedures, experts, and bureaucracies functioning that are authorized by science. They will not disappear, as if by a miracle, if we recover the taste that was cultivated in the eighteenth-century salons: the taste to be interested in science and technology, which also means, since the two are indissociable, the freedom to laugh at them. And yet to relearn how to laugh is never insignificant. How much time and energy is lost today by those who have reason to struggle, to charge at the red rags being waved under their noses with the names of "scientific rationality" or "objectivity"? The laughter of someone who has to be impressed always complicates the life of power. It is always power that is dissimulated behind objectivity or rationality when the latter becomes the argument of authority.

But above all it is the quality of the laughter that interests me. I do not want a mocking laughter, or a laughter of derision, an irony that always and without risk recognizes the same thing beyond the differences. I would like to make possible the laughter of humor, which comprehends and appreciates without waiting for salvation, and can refuse without letting itself terrorize. I would like to make possible a laughter that does not exist at the expense of scientists, but one that could, ideally, be shared with them.

Briefly outlined, this is the problematic landscape within which this book is inscribed. I claim neither to demonstrate, with many references, nor to describe in an objective, complete, exhaustive manner. I will often proceed via case studies, but here these cases have the status of "figure cases" [*cas de figure*], as one says in mathematics: they are not there to prove, but to explore the possibilities of using the political register to describe the sciences, without excluding myself from this register, that is to say, in the knowledge that the "sentiment of truth" is by no means an excuse to not take into account the consequences of what one believes to be true.

T   W   O

Science and Nonscience

In the Name of Science

IN THE *Science Question in Feminism*, Sandra Harding opposes the "empiricist" critique of the sciences to the "radical" critique, a perspective that should be able to set us on the path of laughter: "Can it be that feminism and similarly estranged [*minoritaire*] inquiries are the true heirs of the creation of Copernicus, Galileo, and Newton? And that this is true even as feminism and other movements of the alienated undermine the epistemology that Hume, Locke, Descartes, and Kant developed to justify to their cultures the new kinds of knowledge that modern science produces?"[1]

With "Hume, Locke, Descartes, Kant," and so many others, we are dealing with those theoreticians of knowledge that epistemology traditionally takes as its starting point. With them, scientific practice tries to express its "objective" practice, in principle generalizable to every field of positive knowledge: "the same scientist" could extend "the same type of objectivity" to whatever he directs his attention toward. Against the "methodological and ontological continuum" that takes theoretico-experimental practices as its model, Sandra Harding invokes another continuum, that of the ethical, political, and historical lucidity required of scientists by the science they practice: "A maximally objective science, natural or social, will be one that includes a self-conscious and critical examination of the relationship between the social experience of its creators and the kinds of cognitive structures favored in its inquiry."[2] From this perspective, the experimental sciences no longer represent the entirety of the scientific field. The "cognitive structures" it privileges in fact correspond to a very specific "social experience," that of the laboratory, and on this point the two are so interdependent, as we will see, that the inclusion of a "conscious and critical" examination of their relation is more difficult here than anywhere else. This is why Harding can see herself as a descendant of Copernicus, Galileo, and Newton, while refusing to take them as models, insisting that their true heirs are those (such as feminists and other minoritarian movements) who refuse to extend "outside the laboratory," in the name of science, the norms of objectivity to which the laboratory gives meaning.

"Hume, Locke, Descartes, Kant" obviously do not explain anything in themselves. The image, which they ground in philosophical terms, of an objective scientific procedure, addressing itself to a world in principle subjected to its requirements, would have no pertinence had it not encountered a large number of protagonists with little interest in philosophy but a very great interest in the benefits of the label of "scientificity," which procures a resemblance to this image. Whether the latter refers to God or the theory of knowledge, to epistemology or transcendental philosophy, to operative reasons or the constitutive conditions of scientific progress, it is the consequence that counts: the scientist is transformed into the accredited representative of a procedure, in relation to which every form of resistance could be said to be obscurantist or irrational.

The interest of scientists explains nothing in itself, however, as long as it is isolated from other interests that are also focused on the making available of the world, that is to say, on the disqualification of everything that seems to pose an obstacle to them. We will return to this problem. Let us first of all pause at the problem posed by the coexistence, within contemporary science, of the practices that Harding's criteria allow us to differentiate, even though they all lay claim to the same model of objectivity: creative experimental practices (I am thinking of the deciphering of the genetic code in the 1960s), practices centered on the power of an instrument (such as those focused on the brain, for which the development of ever more sophisticated instrumental technologies allows us to accumulate data we will one day understand), and practices that straightforwardly mime experimentation, with the systematic production of beings constrained to "obey" the apparatus that will allow them to be quantified (such as the all-too-famous rats and pigeons of the experimental psychology laboratories). "In the name of science," innumerable animals have been vivisected, decerebrated, and tortured in order to produce "objective" data. "In the name of science," Stanley Milgram has taken on the responsibility of "repeating" an experiment already realized by human history, and has shown that torturers could be fabricated "in the name of science" just as others have done so "in the name of the state" or "in the name of the good of the human species."

Obviously, I will have to define what I mean by "creative experimental practices." But I can already characterize the slippage of meaning affecting the term *scientific "objectivity"* in the different cases cited. Already, the accumulation of sophisticated instrumental data requires a specific social experiment, one *that is not capable of creating itself,* for this experiment is constructed on the belief in a model of unique progress: all science would begin in an empirical manner; then, through "maturation," it would acquire the mode of production proper to its forebears. The epistemological image guarantees, here, that one day intelligibility will give birth to data; a paradigm or a theory will come along to recompense the empirical effort. When the data itself is relative to an apparatus that unilaterally "creates" the possibility to subject anything and anyone at all to quantitative measurements, the very meaning of the operation presupposes another definition, which is that of science: what it permits, what it forbids, what it authorizes to mutilate. Finally, when "in the name of science" an experimenter reproduces the conditions under which humans have obeyed the orders that have created executioners, he or she demonstrates the existence of a social experiment in which, in the name of science, the different significations of the terms *to obey* and *to be submitted to* can be confused. "In the name of science," Milgram's subjects obeyed the orders that turned them into torturers. "In the name of science," Milgram submitted them to an apparatus that put himself in the position of Himmler or Eichmann.

Last figure case [*cas de figure*]: the one in which cognitive structures privileged by scientists, far from being reflected in a conscious or critical manner, claim to impose an "each" on everyone, that is to say, where the public, defined as "nonscientific," is called on to make common cause with the interests of scientific rationality. This is the case, for example, in the conflict that opposes official, so-called scientific, medicine to the "soft" or parallel medicines.

It is not by chance that medicine is one of those places where the loop closes back on itself, where the public is exhorted to adhere to the values of science. As opposed to other so-called scientific practices, medicine is supposed to have been pursuing the "same" aim—healing—since the beginning of time, and the question of knowing who has the right to practice medicine is itself more ancient than the reference to science. The conflict between "licensed" doctors and those denounced as charlatans, which is indissociable from the "social experiment" of medicine, was not created "in the name of science," but the reference to science has given it a new turn. The stakes of this reference, on a terrain that has always immediately associated practitioners and public — since the target of the denunciation of charlatans has always been the "fooled public" — are all the more interesting insofar as no one, here, would be tempted to "relativize" the difference between doctors of the seventeenth century, for example, and those we go and see today. "Scientific medicine" has effectively hollowed out a difference, whose meaning we can evaluate.

At what point does the reference to science transform the conflict between "doctors" and "charlatans"? I will here put forward the hypothesis that what allows medicine to lay claim to the title of science is not this or that medical innovation, but rather the way it diagnosed the power of the charlatan and explained the reasons to disqualify his power. According to this hypothesis, "scientific medicine" would begin when doctors "discover" that not all cures are equally valid. A cure as such proves nothing; a popular cure-all or a few magnetic passes can have an effect, but they do not qualify as a cause.[3] The charlatan is thereby disqualified as someone who takes this effect as a proof.

This definition of the difference between "rational" medicine and charlatanism is important. It has given rise to a set of practices for testing medications that is grounded on a comparison with "placebo effects." However, it has had the particularity of transforming a singular feature of the living body — namely, its ability to be healed for "bad reasons" — into an obstacle. It implies that scientific medical practice, far from staging the singularity of what medicine deals with (in order to comprehend it), tries to understand how the sick body could, despite everything, assess the difference between the "true remedy" and the "fictive remedy." It therefore takes as a parasitical and annoying effect something that distinguishes a living body from an experimental system, namely, the singularity of making a fiction "true," that is to say, efficacious. "In the name of science," identified with the experimental model, the "cognitive structures" privileged by medical inquiry, whether with regard to research or the formation of therapists, are then determined by the "social experiment" of a practice defining itself against the charlatans, that is to say, also against the power (to which charlatans testify) that fiction seems to have over the body.

When scientific medicine asks the public to share its values, it is asking the public to resist the temptation to be cured "for bad reasons," and in particular to learn how to tell the difference between nonreproducible cures, which depend on persons and circumstances, and cures produced by proven means, which are active and efficacious *for anyone at all*, at least statistically. But why would ill people, who are interested only in their own cure, accept this distinction? They are not "anyone at all," anonymous members of a statistical sample. What does it matter to them if the cure, or the change for the better, of which they will be the possible beneficiary, constitutes neither a proof nor an illustration of the efficacy of the treatment to which they have submitted themselves?

The living body, sensitive to hypnotizers, charlatans, and other placebo effects, is an obstacle to the experimental method, which requires the creation of bodies capable of bearing witness to the difference between "true causes" and anecdotal appearances. Medicine, which derives its legitimacy from the theoretico-experimental model, tends to see this obstacle as something that "still" resists it, but that will one day be subjugated. The effective functioning of medicine, defined by a network of administrative, managerial, industrial, and professional constraints, systematically privileges heavy technical and pharmaceutical investment, the alleged vector of the future on which the obstacle will be subjugated. The doctor, who does not want to resemble a charlatan, experiences the thaumaturgical dimension of his activity with uneasiness. The patient, accused of irrationality, enjoined to be cured for "good" reasons, hesitates. Where is the "objectivity" in this entanglement of problems, interests, constraints, fears, and images? The argument "in the name of science" is found everywhere, but its meaning is continually changing.

Break or Demarcation?

The definition of "science" is never neutral, for from the beginning of modern science, the title of science has conferred certain rights and duties on those who call themselves "scientific." Every definition excludes and includes, justifies or puts in question, creates or prohibits a model. From this viewpoint, strategies that seek definitions through a break or through the search for a criterion of demarcation distinguish themselves in a completely interesting manner. A "break" proceeds by establishing a contrast between the "before" and the "after," which disqualifies the "before." The quest for a criterion of demarcation seeks to qualify the legitimate claimants to the title of science in a positive manner.

The term *epistemological break* comes from Gaston Bachelard, but its extraordinary influence in French epistemology appears to be linked less to the specific content Bachelard constructed for it through examples drawn from physics or chemistry than to the strategic function it played in domains he himself did not tackle. Having become a "cut," it allowed Louis Althusser to sanction the scientific character of Marxist theory. Today it still permits one to posit the institution of "Freudian rationality" as a point of no return, whatever vulgar problems may still be posed by the cure.[4] From this strategic viewpoint, it is possible to affirm (*cum grano salis*, given the intentions and distinction of the authors) that the definition of science in terms of its break with what preceded it has entered into the field of the "positivist" definitions of science.

Where, in this perspective, can we recognize a positive definition of science? In that it proceeds above all through a disqualification of the "nonscience" that it succeeds. For Gaston Bachelard, this disqualification is associated with the notion of "opinion," which "thinks badly," "does not think," "*translates* the needs of understanding."[5] Thus science is always constituted "against" the obstacle constituted by opinion, an obstacle Bachelard defined as a quasi-anthropological given. This struggle of science against opinion becomes, in its most lyrical moments, the confrontation between the "interests of life" (to which opinion is subordinated) and the "interests of the mind" (the vectors of science). In this sense, Bachelard is closer to the "great positivism" associated with Auguste Comte than to the epistemological positivism associated with the Vienna Circle. For the "Viennese," such as Moritz Schlick, Philip Frank, or Rudolf Carnap, the distinction between "science" and "nonscience" does not have the fascinating allure of a creative revolt of the mind against the subjection of life. Rather, it advocates a purification, an elimination of any proposition that lacks empirical content, which implies, first and foremost, the elimination of any "metaphysical" proposition that cannot be deduced from the facts by a legitimate logical procedure.

My "definition" of positivism thus incorporates thoughts that are not only heterogeneous, but explicitly opposed with regard to their objectives. Whereas the theoreticians of the Vienna Circle were seeking a definition of science that was also a promise of the unification of the sciences, all subjected to criteria that are valid independent of their field of application, Gaston Bachelard celebrates the conceptual mutations associated with the work of "geniuses," who are both inventors and illustrations of the difference between science and opinion. However, the common point that my definition makes explicit—the disqualification of what is not recognized as scientific—has the merit of bringing to light not only the truth of authors, but the strategic resources they offer to those for whom the title of science is at stake. From this point of view, the "break," whether it is of the order of a purification or a mutation, creates a radical asymmetry that deprives what "science" constitutes itself *against* of any possibility of contesting its legitimacy or pertinence.[6]

This asymmetry, which characterizes what I am calling positivism, is what permits me to suggest that the difference between this way of characterizing the sciences and their denunciation as "technoscience" is not very great. Above all, it is subject to an inversion. What positivism disqualifies can also be described as the object of an irremediable loss, the victim of a destruction of signification and value. Another typical trait of this asymmetry is that the characterization of "nonscience" is much more clear and more assured than that of "science." Bachelard emphasizes the fact that the "historical" history of the sciences is permeated by opinion, or, in Althusser's terms, by ideology. The problem is that the image of a "slowed-down and hesitant" history, ceaselessly delayed by the "effective pressure of popular science that realizes...all errors,"[7] presupposes a morality that the history of the sciences does not manifest, namely, the separable, because not fecund, character of the error or ideology, which would then denounce themselves. If we think that, by definition, an "ideological claim" cannot make history in the properly scientific sense, we will quickly come to the point of having to slice off entire portions of science that are perfectly recognized today.[8]

The fact that the denunciation of nonscience, as opinion, is more assured in Bachelard than the definition of science has very serious consequences. The disqualification of opinion forbids us from opposing to the definition a science that gives its "object" anything to which this object, thus defined, gives no meaning, or which it denies. For then it is "opinion," which is interested in what the object denies, that would be made to bear witness against science. At the limit, this denial, in itself, can "prove science": science can demonstrate that it has made a break by daring to neglect what everyone was interested in "beforehand." The more the work of mourning for the required past appears tedious and crippling, the more efficacious the theme of the break becomes.

What is interesting about the demarcationist tradition, whose origin is associated with the name of Karl Popper, is that it takes as its point of departure the critique of positivism (in its logical form developed in Vienna). It does so on two points. On the one hand, Popper does not accept the assimilation of nonscientific propositions to propositions devoid of meaning. For him, "metaphysical" questions do not belong to a disqualified past, but express a search for meaning for which the sciences can provide no substitute. On the other hand, the Viennese definition of scientific propositions is too large. It grants the title of science to claimants that Popper deems to be illegitimate. For Popper, these claimants were first and foremost Marxism and psychoanalysis. But for some contemporary epistemologists, such as Alan Chalmers, they can be found in a proliferating population of academic undertakings, from the sciences of communication to administrative sciences, from economics to the pedagogical sciences, which seek in facts, measurements, logic, and statistical correlations the guarantee that they are indeed sciences.[9] It is from this perspective that I will here interest myself in the demarcationist tradition. I will not linger over Popper's "political" theses on the "open society" or his opinions concerning the social sciences. I will focus on the imperative he first puts forward in *The Logic of Scientific Discovery* (1934): we must clarify the difference between "Einstein" and an illegitimate candidate to the title of science.

The fact that Popper takes Einstein as a "scientific type" does not only stem from the renown of relativity, which impassioned the young philosopher. Einstein also expresses the failure of Viennese positivism. The latter had accorded itself two tutelary figures, Ernst Mach and Albert Einstein. Einstein, through his elimination of absolute space and time, seemed to confirm Mach's theses concerning the necessity of purifying science of all metaphysical presuppositions. During the 1920s, however, Einstein broke the alliance that had been imposed on him. He called Mach a "deplorable philosopher," denying any influence, in the fecund sense of this term: Mach's philosophy is good only for "killing vermin." And he confessed to a truly metaphysical motive: the passionate search for a true access to reality.[10] Einstein, who will always be the "true scientist" for Popper, thus explicitly put the positivist reading of science in question.

Thus, for me, what is interesting about the search for a criterion of demarcation between science and nonscience boils down to its attempt to provide a "positive" definition of "true" science. As we shall see, the fact that this attempt might end in failure does not point to the impertinence of the question, which is essential for resisting what is advanced "in the name of science," but to the problem of finding the means of putting it to work. In this sense, this failure — as opposed to strategies that disqualify what a science, in order to impose itself, has already vanquished — will itself be instructive.

Popper's Question

All too often, what is retained from *The Logic of Scientific Discovery* is Popper's "falsificationist" position: whereas no accumulation of facts, however large, is enough to confirm a universal proposition, a single fact is enough to refute (falsify) such a proposition. His adversaries have attributed to him an ambition to establish a methodology for the sciences grounded in this position. Moreover, his student Imre Lakatos has suggested that we distinguish between "three" Poppers: $Popper_0$, the "dogmatic" or "naturalist" falsificationist, who would have had this ambition but never wrote a single line; $Popper_1$, the "naive" falsificationist of 1920; and $Popper_2$, the "sophisticated" falsificationist, which the true Popper had never really been, but which he, Lakatos, had need of in order to arrive at his own solution.[11]

The "triple Popper," the result of Lakatos's rational reconstruction, is not a sign of the complexity of Popper's thought, which has always been perfectly explicit. Rather, it signals a tension, characteristic of this position, with regard to the scope and power of the sought-after "criterion of demarcation." Certainly, it must make clear a difference, but does it have to guarantee the possibility that the entirety of science will respect this difference? If this were the case, the definition of the difference between science and nonscience could engender a "methodological" definition of the productive procedure of science. This is the position attributed to Popper$_0$, and it leads to a variant of positivism, because any procedure infringing on the criterion would thereby find itself disqualified. But if this is not the case, then what does the possibility for a field of research to become "scientific" depend on? The position philosophers take in relation to the sciences depends on this equivocality. Should they abandon any claim to judge or produce norms, which would allow them to tell the scientist "you should have...," so as to ally themselves with "art critics," who know that they do not have to give lessons to artists but devote themselves to commentary, for nonartists, on the singularity of the artistic work?

Popper always adopted a position rather close to that of the "art critic," for above all he "loved" science as Einstein symbolized it for him. The invariant of his career was always: whatever the criterion, it must allow us to understand why Einstein is a scientist and why the Marxists and psychoanalysts are not. His students sought to construct norms that could, if not explain science, at least demonstrate that scientists must subject themselves to certain constraints that would permit their rationality to be verified. By everyone's account, the starting point of the tradition, *The Logic of Scientific Discovery*, published in 1934, is resolutely "antinaturalist": science does not adhere to a "natural" definition of rationality. Popper, after establishing the logical difference between confirmation and refutation, in effect shows that, once one distances oneself from the logical universe in which propositions are defined in a univocal manner, this difference is insufficient. Logic will never be sufficient to impose the conclusion that a proposition has been refuted by an observation—something that Pierre Duhem had already explained in 1906 in *The Aim and Structure of Physical Theory*.[12] In effect, no observation can be stated without having recourse to a language that confers a signification on it and that permits its confrontation with the theory—today we say that every fact is "impregnated" with theory. Scientists are thus perfectly free to annul a possible contradiction between observation and theory. They can redefine the theoretical terms, or else introduce new conditions of application of either the theory or the instrument producing the annoying "fact." In Popper's vocabulary, they can "immunize their theory" thanks to a "conventionalist stratagem." In itself, this phrase expresses Popper's judgment against the "conventionalist" interpretation of science associated with Henri Poincaré, Einstein's adversary. If all our scientific definitions were only conventions, which could therefore be modified at will, Einstein would never have been able to triumph over Lorentz's rival interpretation, which Poincaré supported. Consequently, the demarcation insists on a *refusal* of the freedom that logic leaves to the scientist: truly scientific people are those who know how to renounce the free redefinition of "basic statements" (which permit the statement of observation) and accept the need to deliberately expose their theory to the test of the thus-stabilized facts.

The asymmetry between confirmation and falsification thus engenders no logical rule. Rather, for Popper, it has the status of an *occasion* for an *ethic*. Scientists are scientific because they exploit this asymmetry—something logic does not constrain them to do, but which they can *decide* to do. This decision finds its meaning in the "aim" of science: the production of the *new*, new experiments, new arguments, new theories. Those who, like Marxists or psychoanalysts, profit from the relation of force that will always allow them to interpret the facts in a way that leaves their theory intact, according to Popper, will be logically irreproachable, but they will never produce a new idea. Those who, like the Popperian Einstein, choose to expose themselves to refutation will take the only path open to the search for *truth*, which Popper thus associates with an aesthetic of risk and audacity. In relation to the "aim" of science, our subjective convictions and our search for certainties are defined as venerated *idols*, as obstacles.

In 1934, then, there is no Popperian theory of science, but rather a characterization of the scientist that is, one could say, at once ethical, aesthetic, and ethological. The question is not "How can one be scientific?" but "How can scientists be recognized?" By what passions can they be distinguished? What engagement, which nothing imposes on them rationally, gives value to their quest? What expectations characterize the way they address themselves to the facts? In short, what is their "practice" (in the sense in which this term unites what Kant meant to distinguish in the *Critique of Pure Reason* and the *Critique of Practical Reason*)?[13] What brings the Popperian scientist into existence is not a truth it would be possible to possess, in return for a respect for certain rules, but rather the truth as an *aim*, authenticated by a *manner of relating to the world*, exposing oneself to its challenges, accepting the possibility that it will disappoint our anticipations.

There are many questions to be asked regarding this Popperian characterization. The first, which was asked neither by Popper nor by the demarcationist tradition, is the question of knowing what this characterization is in fact aiming at: the scientist in general, or the specialist in the experimental sciences? For, as Alan Chalmers recognizes, for instance, the set of examples discussed by the demarcationist school is drawn from physics and chemistry, and Popper himself was interested in history and the social sciences primarily in order to criticize historicist, dialectical, and hermeneutic theories, among others, but he never found an equivalent to "Einstein" in such fields.[14] However, even with regard to sciences whose experimental nature is incontestable, we can ask what meaning the criterion of demarcation would have. Is it a "realist" criterion, which would claim to characterize the norms to which true scientists in fact conform themselves? Is this criterion enough to define the activity of the scientist? Does it allow us to understand the history of the sciences we are tempted to recognize as "truly scientific"? This is the question pursued by Popper's principal student, Imre Lakatos.

Popper himself came to recognize rather quickly that, if there is no *fact* that constitutes "progress," then the fact that scientists manage to produce theories that resist falsification during a period of time, and to replace falsified theories by "better" theories that successfully foresee new effects, means that the practice of falsification would tend to make the history of science a rather disheartening cemetery of theories. The latter, as Popper has written, would certainly have managed to prove their scientific character by having themselves refuted, but the doleful repetition of this proof does not constitute a very exalting perspective. The heroism of scientists who accept the need to "expose" their theories certainly implies the acceptance of a risk, but not the resignation of permanent refutation. To be a "true" scientist, according to Popper, one must belong to a field that gives scientists reason to hope that their theory will resist, a field in which the possibility of "progress" is seen as something acquired. But the analysis then becomes tautological. If the condition that allows scientists to conduct themselves in this manner is nothing other than progress, we cannot explain the "progressive" character of the sciences by the behavior of scientists, by the possibility they have to learn and to produce the new. This is what we have to understand.

As we shall see, with regard to the sciences, Popper himself came to adopt a perspective that affirms this tautology in its most radical mode, and gave it a "cosmological" meaning. The singularity of the sciences in relation to the psychological quest for certitudes and confirmations must not be explained by a psychology of the scientist. Like the appearance of life out of material processes, its singularity must be emphasized, and it is this singularity that explains the subjective difference between Einstein and the Marxist or the psychoanalyst. By contrast, the demarcationist school sought to construct a "better criterion" that could claim to describe in a normative manner the constraints to which scientific rationality, even in physics, is subjected "outside tautology."

The Unfindable Criterion

The singularity of the demarcationist tradition stemming from Popper can be located in the way it made use of the history of the sciences, which was seen as a kind of a "testing ground" for the different criteria of demarcation that were proposed. According to Lakatos, whom I am here accepting as a guide, these criteria must allow for a *rational reconstruction* of this history that clearly establishes the difference between the anecdotal dimension and progress. Thus, a criterion that disqualifies a position we judge to be useful and necessary to scientific progress would not pass the test of history. And the first victim of this test was Popper's "heroic falsification."

What would have happened if Copernicus had been a heroic falsificationist? A disaster, for he would have heroically abandoned his heliocentric position, which was refuted by the fact that the theory implied that Venus had phases like the Moon, something that astronomers had never observed. As Lakatos says, every theory "is born refuted." To have a chance of succeeding, it needs to be protected and cherished by its promoters. One might then try to define a "sophisticated falsificationism," centered on the notion of progress. The judgments of scientists would henceforth have to be oriented around the possibility of confirming audacious conjectures, such as the heliocentric theory, or falsifying prudent conjectures, those derived from a knowledge that can be considered as acquired. The first consequence of this position is that the judgment of rationality must be made according to the references of the epoch, which defines audacity as much as acquired knowledge.

However, falsificationism, whether naive or sophisticated, remains centered on a typical "scene," the confrontation between a theoretical position and an observation. This scene is directly inspired by a positivism of the logicist type, which reduces science to a double source of knowledge: *facts* that are observable and particular, and a *reasoning* (whether of the inductive or falsificationist type) that constructs a general theoretical proposition from the facts. But, Lakatos protests, the history of the sciences manifests such scenes only through an artificial a posteriori reconstruction. The "crucial experiment," in which the scientist deliberately exposes his or her theory to experimental proof, is probably the most rhetorical and artificial scene of history. More often than not, it is *after* the experiment, *once it has succeeded*, that it is put on stage as having been crucial; and it in fact constitutes the public and highly ritualized putting to death of a rival hypothesis.

In other words, it is not enough to say that facts are "impregnated with theory" and thus can be reinterpreted at will. This way of presenting things tends to transform itself into a difficulty, an obstacle to the "primordial scene," that of the confrontation between fact and theory, which is, according to Lakatos, the very material of the history of the sciences. Historically, an observable fact is not confronted with a proposition, which it verifies or refutes; it finds its meaning in a *research program*.

Like "sophisticated falsificationism," which implies that "audacious conjectures" can be verified, the notion of the research program, we must emphasize, presupposes the success of the sciences it characterizes. In effect, this notion expresses a *differentiation* that would have no meaning if a theory were content to "survive" without creating the conviction that it indeed constitutes a path of privileged access to the phenomena it concerns: the differentiation between the "hard core," to which this privilege will be linked, and the "protective belt," where there will be ceaseless negotiations between the significations relative to the "facts" and the statements with which the hard core is concerned.

In the dynamic viewpoint instituted by the research program, there is therefore no confrontation between a fact and the research program as such, for in itself the fact is never capable of putting the core of the program in question. The confrontation takes place only between theories that belong to the "belt," theories that can be modified in numerous ways while still confirming the veracity of the core. Within a program, the mode of negotiation thus emerges naturally from the "conventionalist stratagems" that Popper denounced, immunizing the core from any refutation by the facts. Scientists do not have to "decide," according to dogmatic, naive, or sophisticated criteria, whether or not there has been a refutation. Within their research program, they must "accommodate" the facts with this or that part of the protective belt in a way that reestablishes the coherence of the whole. But where, then, does the demarcation lie, that is, the difference between a truly scientific program and "false science"? The decisive site, for Lakatos, lies in the evaluation of the mode of transformation of the program in the long run: Is it progressive or degenerating? Scientists do not have to make an *instantaneous decision*, as in the scene of confrontation, but they must ask themselves if the modifications brought about in the protective belt of their program have, in the course of time, accrued for it a predictive power, have given it access to new types of facts, have been testable independent of their function of accommodation—or if, on the contrary, the program is constantly weighed down by ad hoc accommodations, accommodations in which no other signification can be recognized other than that of having protected the hard core. If he concludes that his program is degenerating, the rational scientist, in a progressive phase, will abandon it for another program.

Lakatos thus preserves the necessity for a decision, and above all the definition of criteria that allow scientists to judge the decision they are making, in this case, whether or not to abandon a program. It is here, in fact, that the demarcationist tradition recognizes its own: whoever states an imperative of decision states the possibility of evaluating "true" scientists by their lucidity, by the critical relation they maintain with their own activity. True scientists *are not subjected* to a norm, as is the case with Kuhn's normal scientist; they *subject themselves* to a norm, thereby assuring that science escapes a sociopsychological description and is subjected to a theory of rationality. In order to guarantee the possibility of judging, however, the norm must be *explainable*. And it is here that Lakatos's research programs in turn encounter the test of history. Just before his death, Lakatos himself wound up recognizing that the judgment of the "man of science" could only take place retroactively.[15] It is we who now know that this or that program was degenerating. But in this case, it is history itself that gives philosophers the power to judge, to determine "at what point" it was rational to abandon this program for another. And this power, conferred by history, is in fact *redundant*: philosophers confirm the "losers" that they have indeed lost, but they have no proper resource for evaluating and judging the reason for which these losers hung on to their program; they can only say that history has not retained these reasons.

Lakatos's conceptions encounter other difficulties, which I will not linger over here. They imply, notably, that the normal situation in science is the competition between rival research programs—which is what allows scientists to exercise their critical capacities. Here, the historical style of Lakatos and his disciples comes into conflict with the style of Kuhn and his disciples, who emphasize the solidarity between the "crisis" that traverses a program and the invention of an alternative program. But the most important point, which in my eyes marks the end of the demarcationist tradition, remains the impossibility of formulating in explicit terms a set of criteria that, while informed by the past, would be valid for the present. In other words, it does not provide an explanation of the rationality at work in science, but rather of the history that gives the philosopher of science the power to judge, and this only to the degree that that history can be read, in physics as in chemistry, through the mode of progress. The demarcationist tradition, far from explaining the progress that recompenses "true" science, winds up leading to a commentary on the way the "true sciences" have progressed.

One Historical Tradition among Others?

Reason: there are many possible readings of this word that haunts philosophy. We could justly say that one of the poorest readings is that of normative rationality: the search for a criterion to which anyone who wants to be a scientist must agree to subject himself or herself. But its importance lies in the fact that it was born out of the concern to demonstrate that science is indeed irreducible to the registers we are accustomed to use in deciphering human activities—that is, to demonstrate in an explicit manner what scientists themselves affirm about science.

The failure of this formulation, moreover, stems from this very concern. Its failure does not threaten thinkers who isolate, in scientific production, a moment or a work in which the labor of "reason," as they conceive it, has allowed itself to be apprehended. Such readings of science must be termed *edifying* to the degree that, just as the lives of the saints illustrate the power of grace, the life of the sciences or its concepts can illustrate an idea of reason. Philosophers attribute to themselves the right and duty to identify certain conceptual mutations in the sciences that they judge, rightly or wrongly, to be significant, and to construct a philosophical characterization of reason on this basis. Against this undoubtedly exalting vision, I have the weakness of preferring a more vulnerable approach to history, one that will allow us—despite the power to judge that we have conferred on ourselves, the heirs of the judgments of this history—to speak of "failure."

Nonetheless, what are we to make of this failure? What are we to make of the impossibility of formulating criteria that would validate science in a general manner, and thus grounding the possibility of a discourse on science that would differentiate science from what merely resembles it? Should we conclude, with Paul Feyerabend, a disenchanted disciple of Popper, that any claim to define "the" difference is only propaganda?

In *Against Method*, Feyerabend clashed with established sentiments by comparing scientific activity with astrology, voodoo, and indeed the Mafia, and he paid a price for this strategy: those he was clashing with reduced the problem he posed to this scandalous comparison.[16] But what was at stake in Feyerabend's "relativist" position was not the assimilation of Einstein to an astrologer or Galileo to a mafioso. He was trying to show that, in order to succeed in making history, in making us accept what he proposes as "objective" knowledge, a scientist cannot be content with what the philosopher thinks of as "objective." The construction of objectivity has nothing objective about it: it involves a singular but not exemplary manner of relating to things and to others, like a mafioso activity.[17] But this does not mean that it originates in the same type of engagement as does mafioso activity.

Feyerabend's thesis is thus not directed against scientific practice, but against assimilating objectivity to the product of an objective procedure.[18] Despite its apparent character as a truism, this assimilation is in fact an instrument of formidable power. It makes objectivity the general destiny of our knowledges, the ideal at which they must aim. Every practice of knowledge will be called on to differentiate the things it tends to confuse if it is not scientific: objective, scientific knowledge on the one hand, and projects, values, significations, and intentionality on the other.

In this sense, Feyerabend's first target is positivism as I have defined it, including its denunciative variant, insofar as the latter assimilates the advance of "technoscience" to a destiny determined by its inexorable identity, stronger than the (good) intentions of scientists. Also included among his targets is the marvelously scientistic discourse held by so many theoreticians of human subjectivity who hand over to objective science everything that is not "the subject," its rights, its values, its freedom, and so on. There is nothing neutral about this gesture: to render unto Caesar what is Caesar's is also to claim for oneself everything that does not belong to him. The generalizable triumph of objectivity, recognized in principle, depends on the possibility of setting oneself up as the representative of subjectivity as such, which is then recognized as *the other pole*, indestructible and inalienable, of the human mode of existence.

It is against this distribution, in which brothers who are apparently enemies are in fact thick as thieves, that Feyerabend writes: "Decisions concerning the value and the use of science are not scientific decisions; they are what one might call 'existential' decisions; they are decisions to live, think, feel, behave in a certain way."[19] In other words, objectivity, when it is produced, in no way permits one to designate subjectivity as its other pole, finally purified and free to define itself. Thus defined, the "subjective moment"[20] is merely a "remainder," the product of the forgetting of the "decision" that produced objectivity, and of its consequences for the way we "live, think, feel, and behave."

Nonetheless, Feyerabend's strategy, to the degree that it is rooted in a failure—that of the formulation of general criteria of scientificity—has its weaknesses. It effectively destroys the relation of belief in objectivity, but the thesis according to which "there exist no 'objective' reasons for preferring science and Western rationalism to other traditions,"[21] salubrious though it may be, is a rather abstract solution to the problem of the "great division" separating our societies, which have produced "science," from all others. Certainly, to the question posed by

Feyerabend with regard to nonscientific traditions—"Were they eliminated on rational grounds, by letting them compete with science in an impartial and controlled way, or was their disappearance the result of military (political, economic, etc.) pressures?"[22]—it is difficult to respond otherwise than he does, but the alternative is no more pertinent. Is the fact that "Western science has now infected the whole world like an infectious disease"[23] entirely determined by military, economic, and political relations of force? Does it owe nothing to the sciences themselves? Is not Feyerabend the relativist still too rationalist when he presents an "impartial and controlled competition" as the only arena in which the sciences could legitimate the proper role they have played in the triumph over other traditions? In other words, the thesis according to which science constitutes one historical tradition among others is vulnerable when it is translated into reductionist terms: science only constitutes one historical tradition among others, the only "true" differences being external factors—political, military, economic. A strategy of unveiling and denunciation.

The first book signed by Feyerabend the "relativist," *Against Method*, was dedicated to Imre Lakatos, "friend and brother in anarchism." It was Lakatos's failure to construct a demarcation, as well as the lucid honesty with which Lakatos recognized his failure, that Feyerabend saw himself inheriting. And the vulnerability of his thesis in relation to his reductionist rendering is also an inheritance of demarcationist epistemology: if science cannot claim any epistemological privilege, it loses any title to affirm its difference from the epistemological point of view. Instead of saying "farewell to reason," Feyerabend could have said "farewell to epistemology." This is what I will do here, conserving from this inquiry the impossibility of understanding the activity of the individual scientist independent of the historical tradition in which his engagement, and perhaps his singularity, is rooted.

T H R E E

The Force of History

The Singularity of the History of the Sciences

SCIENCE OFTEN gives the impression of being an "ahistorical" enterprise. If Beethoven had died at birth, his symphonies would never have seen the light of day. By contrast, if Newton had died when he was fifteen, someone else, in his place... This difference obviously refers in part to the stability of certain problems, in this case the observable regularity of celestial movements, a question that was undoubtedly able to persist for a long time. Moreover, it is not as general a question as one might think. Thus, I believe I can say that if Carnot had died at birth, thermodynamics would not be what it is. But the impression of ahistoricity is nonetheless a singularity of the history of the sciences, which contributes to the explanation of why, until now, it has been rather little frequented by professional historians.

The very existence, some years ago, of a quarrel between "internalist" and "externalist" historians is one symptom of this. What other field could provoke the idea of a division of this sort, between the history proper of scientific productions on the one hand, and, on the other, the history of institutions, of scientists' relations with their milieu, or of the social, economic, and institutional constraints or opportunities that affect the scientific field during a given epoch? We can certainly posit the principle that the sciences *must*, like every other human practice, be situated in history, and that, from this viewpoint, they cannot exist there in a compromised or half-finished manner. But this legitimate ideal does not get at the economy of the problem: Why is it that this situating in history does not go without saying?

It is not enough, here, to invoke the "technical" character of scientific questions, or the fact that historians can allow themselves to be impressed by scientists or epistemologists. These arguments, which lead to solutions of the "it's only" [*il n'y a que*] type, seem to me to mask a much more interesting problem, immediately linked with the conviction that is shared, as we have seen, by many participants in the adventure of the modern sciences: the sciences are not one social practice among others. In other words, the question of the history of the sciences will permit me to take a new approach to the singularity of the sciences, as a putting to the test of the historians' practice.

In a general manner, serious historians will protest if one suspects them of using the passing of time as an instrument of power, allowing them to judge a past situation, to sort out what the people they bring on the scene knew, believed, wanted, or thought. But usually, the discipline they impose on themselves is made easier by the passing of time, which has already allowed them to "equalize" those who, in the past, could believe themselves the victors or to live as the vanquished. In the future that has since come to pass, they have all been subjected to multiple interpretations and reductions. This is what allows historians to construct their own positions—they are the ones who refuse this facility and try to recompose what has been decomposed.

Now the history of sciences brings onto the stage actors whose singularity seems to be, precisely, that they aim at things that the passing of time cannot "make equal." One way of stating the imperative of objectivity—to which any proposition recognized as scientific must correspond, in one way or another—is the following: "That no one, in the present and if possible in the future, would be able to reduce what I am proposing, to distinguish, in my propositions, what is attached to my ideas, to my ambitions, and to things; that no one could identify me as the author in the usual sense of the term." Innovative scientists are not only *subjected* to a history that would define their degrees of freedom; on the contrary, they take the risk of being inscribed in a history and trying to transform it. The actors in the history of the sciences are not humans "in the service of truth," if this truth must be defined by criteria that escape history, but humans "in the service of history," whose problem is to transform history, and to transform it *in such a way that their colleagues, but also those who, after them, will write history, are constrained to speak of their invention as a "discovery" that others could have made.* The truth, then, is what succeeds in making history in accordance with this constraint. To the degree that something an author produces effectively succeeds in making history, this history, far from facilitating the work of the historian, will thereby create a differentiation that becomes ever more difficult to put in question. As for the "vanquished," the historian can even try, in his free time, to make their convictions intelligible; he can also bring to the fore the way the winners were "despite everything" the children of their epoch, by showing the contrast between what they believed they discovered and what science now tells us they discovered. But this contrast itself expresses the power of the discovered truth, because the historian here lets himself be defined by the passing of time, by the difference between what the history *of the sciences* makes him capable of putting in question, and what this history has defined as incontestable.

Thus, in *Études sur Hélène Metzger*, Bernadette Bensaude-Vincent has shown that the "history of ideas and doctrines" style adopted by the historian of science Hélène Metzger in one of her books, *La Chimie*, brutally gives way, for chemistry after 1830, to a pedagogical account of discoveries and theories that follow one upon the other and accumulate.[1] In the same work, Gad Freudenthal read the narrative style that Metzger adopted for chemistry before 1830 as part of the hermeneutical tradition: it was a question of "doing justice" to authors, rehabilitating them, making them interesting, situating them within their epoch by reconstituting their horizon of thought. Does the hermeneutical style of history cease to hold once chemistry becomes "serious" or "truly scientific"? Is there no longer any need to "understand" the chemist? Has the latter become "objective"? Or escaped the air of the time? This was the thesis of Hans-Georg Gadamer, who excluded scientific practices from the hermeneutical field. But this exclusion is in itself a confession, which brings to light the power that historians usually benefit from with regard to their actors—namely, the power conferred by the distance of time.

As Judith Schlanger has noted, in the same set of studies, this situation puts Metzger's style in question at the very points she would be capable of utilizing it. As is often the case when historians of science are inspired by the procedures of art historians, this style tends to overvalue the emergence of a new mode of perception and to undervalue the practices of argumentation. It translates the fact that we are no longer taking seriously the arguments exchanged by the actors of the epoch (since the intervening history has made them outdated ...). For Schlanger, there cannot be a single historical process applicable to the history of philosophy, art, and science, for each of these enterprises is defined by a specific relationship with its own past. In this case, we can conclude that, contrary to what Gadamer thinks, scientific practices and hermeneutical practices nourish a very strict relationship, but in the sense that the former can be defined by its antagonism toward what the latter requires. When the historian "succeeds" in rehabilitating an author by situating him in his epoch, what the historian is conveying is the undoing of this author as a scientist, for he shows that we can henceforth enter his laboratory as if it were a windmill, open to all the influences of the epoch.[2]

Thus, at the heart of the history of the sciences, whether it is inspired by hermeneutics or sociology, there is a difficult relation of force between historians and their actors. And this relation is all the more difficult in that historians themselves have the greatest difficulty in not adhering, if only on the quiet, to the idea that there is indeed progress in the sciences. The asymmetry established between the victors and the vanquished of history is not only an aspect of the situation that historians have to study; it is also an aspect of the heritage that constitutes the historians themselves. How, in fact, could they not think, *like all of us*, that the Earth revolves around the Sun, that microbes are the vectors of epidemics, and that the anti-atomists were wrong to see an irrational speculation in atoms from which chemistry would have to be purified? It is easy for them to put Christopher Columbus in history because Christopher Columbus, by all accounts, did not know that he was going to "discover America." It is difficult for them to recount the work of Jean Perrin, trying to impose atoms on his contemporaries by showing that it was possible to count them, without repeating Perrin's own words, that is, without ratifying the success of what one might call the scientist's "vocation": to *oblige* historians to pass through his own reasons in recounting his work.

"Putting to the test" does not signify an obstacle. The history of the sciences is not an obstacle to the history of the historians, but requires the latter to effectively conform to the "principle of irreduction," refusing to reduce a situation to what the passing of time gives us power to say about it today. The big difference is that this principle, here, is not synonymous with a "methodological decision" that requires the historian to abstain from bringing into play the power conferred by the passing of time. He can, certainly, as did Feyerabend and as do most sociologists of science, attach himself to the indecisive part of a controversy, or to cases where a controversy was not closed in a stable manner.[3] But he should not be surprised to have "hurt the feelings" of those he describes, and who think, for their part, that history should not demonstrate its method in cases where the adversary is weak, but where he shows himself to be strongest (which is what I will try to do with Galileo).

The Three Worlds

Let us approach the question of the "force of history" constructed by scientists from the point of view of its effects on one representative of the epistemological tradition,

Karl Popper. The theory of the "three worlds" developed by Popper in 1968 is both a radical expression of the problem created by the force of this history and a very curious attempt at a solution, which abandons epistemology for a form of generalized philosophy of evolution.

Everything begins, in an apparently trivial manner, with what Popper calls "the principle of transferral." Psychophilosophical theories of the individual acquisition of knowledge, theories of scientific rationality and the collective growth of knowledge, and biological theories of evolution all attempt to characterize a progress, the production of something new and interesting. But how should we characterize this thing that is "produced"? The temptation is obviously to look for a positive foundation that can explain how the new can in fact claim to be "better," that is, that allows us to judge and authenticate the legitimacy of this claim. This is what logicist epistemology tried to do in the sciences: to ground the validity claims of the theories produced, and thus to justify the fact that some are more valid than others. Now, Popper reminds us, logic fails us here, because if we trust it no general proposition could ever discover facts in a valid manner: the procedure of induction, which allows us to pass from a set of particular statements to a general statement, does not allow us to prove this statement, that is, to exclude the possibility that some day or another a fact will come to light to falsify it. Now *what is true in logic is true elsewhere*: this is the principle of transferral. All the modes through which we characterize progress must thus be subjected to the fact that a novelty never founds a positive foundation, which guarantees its (adaptive) value, (psychological) certainty, or (scientific) truth.

The description of the heroic scientist, if it had been adopted as an "explanation" of progress, would already have put epistemology in communication with a psychological theory of apprenticeship through trial and error and with a "mutationist" version of Darwinism through the proliferation and elimination of mutants. The selection eliminates those of whom we can say nothing more than "they were unable to resist the selection." Of the survivors, we can only say "they have not yet been eliminated." The general weakness of this triple theory is that it defines trials, mutants, and theories as indefinitely renewable commodities, which we are never lacking.[4] But when he explicitly introduced the principle of transferral, Popper was already adhering to a nonmutationist version of Darwinian evolution: the success of a living being is not a "survival" but the coinvention of a world of possible resources and a way of relating to this world. In the same way, as Popper notes in *The Unrealized Quest*, infants learn because they are disposed to learn from birth, the success of the innate dispositions to learn implying the human world without which they would have no meaning. In the same way again, scientific theories require a positive characterization: for their refutation to teach us something, it is first of all necessary for them to have had a certain success, for them to have signified a progress in knowledge, the invention of a world they render (partially) intelligible. In all three cases, novelty has no signification independent of the *situation*. The whole must be described, and not judged by criteria more general than the situation itself.

But how to describe a situation? According to Popper, it is in terms of anticipation, which give meaning to the world by selecting and interpreting certain of its traits, and in terms of the risks these anticipations entail. The first term has become the "problem," which creates a new situation (even if the novelty of the problem often cannot be perceived independently of the formulation of a new type of solution). The "problem" is recognized in its capacity to subsist throughout various "attempts at a solution" or "conjectures" (physiological, behavioral, or conscious), and it is this subsistence that allows us to understand the elimination of "erroneous" solutions and the possible creation of new problems. According to Popper's now-omnipresent schema, P1 engenders TT (tentative theory), which engenders EE (elimination of errors), and can engender P2.

A decisive moment is produced here. The subject of the evolution of science is no longer the individual, whether psychological or ethical. The scientist is defined by the situation. Consequently, the ethical prescription is no longer necessary to define science, and the disqualification of the adversary takes place in these new terms: Marxist or psychoanalytic, the adversaries are those who hang on to their hypotheses and refuse the problems posed by their situation in the world. But this disqualification has now become "ontological." Marxists and psychoanalysts, like the amoeba and every other animal, are enclosed in the "second world," that of beliefs and convictions, desires and intentions, whereas the "true" scientist is defined by the emergence of a "third world," that of objective knowledge. The master contrast is displaced, it now bears on the difference between Einstein and the amoeba: the latter is identified with its hypotheses and dies with them, whereas Einstein lets his hypotheses die in his place.

At first sight, the reader might consider Popper's solution to be calamitous, for the difference between science and nonscience—a problem that scientists, after all, do not seem to have much trouble resolving—here implies an *onto-logical* difference between *the second world*, that of living beings with their convictions, their fears, their desires, their intentions, their beliefs, whether conscious or not, whether psychic or incarnated in the organs of perception and their metabolism, and *the third world*, that of objective knowledge. But the reader would be wrong to think that, having made this distinction, Popper once again takes up purely and simply with the tradition of "great positivism," with a few cosmic frescoes presenting humanity's ascent toward reason. This would lose sight of the distinctiveness of path taken by Popper, whose starting point is logic's inability to give an account of scientific knowledge, and the generalization of this inability by the "principle of transferral." The singularity of this starting point is that it poses the problem of "the force of the sciences" by beginning with the question of the pertinence of our anticipations when we want to describe them. Before interrogating the products of a situation, we must first of all recognize the references it itself has produced. Because logic cannot justify science, it is not necessary to conclude that science is illogical, but that, with science, a logic of the situation has emerged in relation to which logic is not pertinent.

The difference between the second world and the first world — the world of material, geological, physicochemical, and meteorological processes — is exemplary in this regard. When we are dealing with a living being, we know that the pertinent mode of description must include the living being's "point of view" on its world, whether this point of view is indissociable from its metabolism, as is the case with the amoeba, or whether it can refer to a psychic dimension, as seems to be the case with mammals. Whether dealing with amoebas, chimpanzees, or ourselves, *we cannot be described without taking into account the fact that not all environments are the same for us.* In other words, the distinction between the first and second world celebrates the *emergence* of beings that can certainly be analyzed in terms of a process belonging to the first world, but that imposes, in order to be understood in a pertinent manner, a new language. It is in this language, notably, that we can with good reason hesitate between "cause" and "reason," that is, we can, without metaphor or anthropomorphic projection, speak of "differences that make a difference," as Gregory Bateson would have said. The second world is the world in which *meaning* emerges.

There are many ways of distinguishing between meaning and signification [*sens et signification*]. One of these ways, which I will adopt here, creates the terrain required by the Popperian distinction between the second and third world: as opposed to meaning, signification implies that the people for whom it is a reference are not surprised to be asked to explain or justify it. This distinction is aesthetic, ethical, and ethological: it characterizes a way of existing in a mode that implies that, if need be, one might "have to give an account" of the way one exists. Signification implies the emergence of a possibility of describing, examining, and discussing which, by vocation, attributes to the interlocutor an anonymous and impersonal position. This possibility corresponds to a new problem, to a logic of the new situation — and often to the institution of a relation of force between those who ask for or look for accounts and those who do not even know they have to give an account. One thinks here of grammarians and other regulators of language in relation to those, like Monsieur Jourdain, for whom speaking is like breathing. But in no case does this correspond to the guarantee that the account given would be capable of grounding its own adequation, that the explanation would be sufficient, coherent, and true.

It goes without saying that, for Popper, everything that is human confuses meaning and signification. But for him, the distinctiveness of science is that, from this "terrain" that constitutes living beings who "seek to give an account" and thus posit themselves as the problem of truth, legitimacy, and certainty, it makes a dynamic emerge that transcends this preoccupation. To give an example, it is possible that the mathematical demonstration invented by the Greeks had been, at the beginning, little more than a way of grounding the certainty of the statement, but the very activity of definition and demonstration led to a completely different history. With "irrational numbers," a scandal for Greek reason, one has the archetypal example of an inhabitant of the *third world*, capable of imposing itself despite the intentions and convictions of the subjects of the second world.

For Popper, the force of the history constructed by scientists is thus linked to the fact that "psychological" subjects are not its masters, but rather are constrained by the problems they make emerge. And correlatively, what this history *imposes on those who want to describe it* is that they must take into consideration the third world and its relative autonomy in relation to subjects endowed with intentions and convictions in their search for certainty. Science celebrates the crossing of a threshold from which *it is impossible not to recognize* that the central actor of evolution is no longer the subject belonging to the second world but the *objective problem* inhabiting the third world. Those who do not recognize this try to ground scientific knowledge in terms of criteria of legitimacy or proof, which corresponds to the search for certainty by the inhabitants of the second world — even if this means that, if they fail, they become relativists like Feyerabend rather than asking if their questions were pertinent.

The articulation between the second and third world thus reproduces the one that prevails between the first and second world. Every problem has as its condition the emergence of activity of a subject (a nonintentional activity relative to the event of emergence), but once it exists, it subsists and provokes those who will subsequently be at its service, those whose intentions, convictions, and projects *can no longer be described* independent of this new type of situation.[5]

I have just presented the Popperian theory of the "three worlds" in the guise of a challenge rather than as a solution. The challenge is a pertinent one. It concerns, in its maximal radicality, the question of power that the distance of time gives the historian in relation to the problems of its actors and their arguments, and puts the singularity of the history of the sciences under the sign of the confrontation between two powers: that of interpretation, which recognizes beliefs, convictions, and ideas everywhere, and that of the problem, whose imperative gives rise to the existence of the scientist.[6] But if this is the challenge, the solution proposed by Popper, for its part, is "impregnated" with the epistemological preoccupations of his starting point. I will here point to three major weaknesses, which at the same time designate three constraints for the construction of the solution I will suggest in what follows.

On the one hand, Popper's staging is undertaken in order to lead to a perspective that conserves the ideal of a pure science and the correlative definition of an "external milieu" as impure, which always risks contaminating scientific purity and putting science in danger. In other words, one of the vocations of the world of Popperian problems is obviously to clear out any political dimension, which Popper would identify without hesitation as the second world. *Can we transform the use of the words* politics *and* scientific problem *radically enough so that their vocation will no longer be to mobilize arguments in the perspective of a confrontation?*

On the other hand, Popper's third world ratifies the privilege of the mathematical and experimental sciences, because it is in these sciences that history or progress seems to refer in the most plausible manner to the problem as a product emerging from human activity, the function of the world being that of submitting to the questions inspired by these problems. The idea that the world could itself pose a problem, in the sense that it could itself become the "central actor" that subsists and provokes those who describe it, is foreign to Popper's theory, but, as we shall see, it can intervene in the question of the difference between the experimental sciences and the field sciences [*sciences de terrain*]. *Can we comprehend the practical differences between the sciences without ratifying their hierarchization?*

Finally, and above all, the three Popperian worlds constitute a perspective that is at once *too vast*, allowing us to create a contrast between Einstein and the amoeba, and *too poor*, remaining silent on the difference between the way a problem, scientific or not, is able to impose its conditions and the way a scientific production is imposed historically, and *too determinist*, giving the problem the power of assessing the difference between those who will be its vectors, and the rest, who will be seen as mere obstacles coming from the second world. *Can we avoid conferring on the problem the power of defining science, that is, of transforming its history into an ontologico-evolutionist model?*

What, then, in the end, should we retain from Popper? That historians of science certainly have no need to feel obliged to recount history the way its actors recount it, but also that they have no need to decide a priori whether what its actors say, when they speak of their own engagement, is mythical, ideological, deceptive, or too tainted with epistemology. A situation — to the degree that it arouses actors who refer explicitly to the constraints it brought into existence — is not reducible to its milieu of emergence, any more than the way of relating to the world that invents a new species is reducible to the constraints that, as we know a priori, must be satisfied: reproducing, finding enough food, having a sporting chance to escape from predators, and so on. This does not mean, of course, that the invention or the situation can be separated from the milieu in which it is produced. It is because he respects this irreducibility that Thomas Kuhn, I believe, was so well understood by scientists, whereas he scandalized epistemologists, including Karl Popper.

Clarifying the Paradigm

The misunderstanding that has surrounded the notion of the "paradigm" introduced by Kuhn results from the reductionist image that assimilates it to a merely professional and institutionalized norm, a purely human convention imposed with dogmatism by hunting down and stifling any lucidity and critical spirit. We could also speak of a "crowd psychology," as Lakatos does, or suggest that a discipline is founded by making reign a rather strict repressive order so as to eliminate the proliferation of rival hypotheses, or affirm that the notion of paradigm saves us once and for all from the care of having to determine how nature has a voice in the subject matter of the sciences: it does not have it here any more than elsewhere. Kuhn, in this sense, would herald and prepare the terrain for Feyerabend.

Kuhn relates how an enthusiastic colleague said to him during a conference, " 'Well, Tom, it seems to me that your biggest problem now is showing in which sense science can be empirical.' My jaw dropped, and still sags slightly. I have total visual recall of that scene and of no other since de Gaulle's entry into Paris in 1944."[7] This imperishable memory conveys the depth of the misunderstanding between the author and those who made use of his work. From the start, Kuhn has played a central role in my staging because of the completely divergent reactions he evoked among epistemological philosophers and scientists. But the satisfaction of scientists does not merely stem from the way Kuhn preserves the autonomy of scientific communities; as we will see, it can also be explained by the intrinsic link he constructs between this autonomy and the impossibility of reducing the paradigm to a sociological or psychological reading.

Whatever Kuhn might be criticized for, there is one thing on which he is perfectly clear: the paradigm cannot be interpreted as a "purely human" decision, whatever decision theory one might like to invoke. No human decision, no constraint, no indoctrination can eliminate the difference between sciences in which a paradigm has "come about" and those in which it has not. This is because a paradigm is not simply a way of "seeing" things or of posing questions and interpreting results. A paradigm, first and foremost, is of a *practical* order.[8] What is transmitted is not a vision of the world but a *way of doing*, a way not only of judging phenomena, of giving them a theoretical signification, but also of *intervening*,[9] of submitting them to unexpected stagings, of exploiting the slightest implied consequence or effect in order to create a new experimentation situation. All these are what Kuhn terms "puzzles." This term means that, during a normal period, the failure to solve a problem of this type will put in question the competence of the scientist and not the pertinence of the paradigm, exactly as in a social game. But the mentality of a "puzzle lover" is created neither by indoctrination nor by the repressive rarefaction of rival "rules of the game." It is not enough, no matter where one looks, to find situations that resemble a model or confirm a theory. It is necessary for the appetite to be sharpened by the challenge — not by a monotonous and unanimous landscape, where one always "recognizes" the same thing, but by an undulating landscape, rich with subtle differences that must be invented, where the term *recognize* does not refer to the observation of a resemblance but to the challenge of actualizing it.

Like Kuhn, Lakatos emphasized the highly artificial character of the logicist mise-en-scène confronting an isolatable proposition and data that either confirm or invalidate it. But his own mise-en-scène, inasmuch as it remained centered on the confrontation between "observable facts" and the "research program" (equipped with its protective belt devoted to a negotiation with the facts) remained equally dependent on logicism. In effect, it inspires the idea of a gathering of facts that can be defined independently of the theory, so that one can then compare and negotiate between the facts and the theory. Against this idea, Kuhn introduced the notion of incommensurability between the empirical reference of rival paradigms. This, of course, caused a scandal among philosophers: Does not the fact that no common language can create the scene of an "impartial and controlled competition" between two theories faced with the *same* facts prove that scientists are fanatically enclosed within their own version of the world? This misunderstanding stems from the fact that the notion of the paradigm corresponds not to a new version of the "impregnation" of facts by theories, but to the notion of *the invention of facts*. To speak of impregnation is to conserve the ideal of a pure fact, gathered as such, and to designate a gap or a lack, surmountable or not, in relation to this ideal. To speak of invention is to abandon this ideal and to affirm that experimental facts are "authorized" by the paradigm, in both senses of this term, referring both to the source of legitimacy and the responsibility of the "author." Facts lose all relation with the idea of a common material whose ideal vocation would be to assure the possibility of comparison or confrontation (the logicist or normative mise-en-scène). Their primary definition is not to be observable but to constitute *active productions of observability*, which require and presuppose the paradigmatic language.[10] This is why, according to Kuhn, two "paradigms" or "research programs" usually do not coexist in such a way that the scientist has to evaluate their respective modes of development. Such a coexistence involves the idea that, in a general manner, facts are preexistent and can nourish one or more programs; it does not recognize their invention. Normal science less explains what preexists it than it creates what it explains.

In short, it is precisely because a paradigm must have the power to invent facts, practically and operationally, that it itself is not invented, or in any case not in the same sense. The invention of facts is competent, discussable, and astute, whereas the "invention" of a paradigm, for Kuhn, is imposed in the manner of an *event*, creating its before and after. A *rare* event, for it constitutes the discovery of a way of learning, saying, and doing that institutes a singular *relation of force* with the corresponding phenomenal field. The tradition of demarcation collides with a *general* problem, that of the power of interpretation, the power possessed by *every* language to bend the facts, to negotiate significations. Kuhn's paradigm designates an event-power: a mode of mobilizing phenomena manifests itself, in an unexpected manner, almost scandalously fecund. Even more than some sort of indoctrination, it is this scandal that nourishes the conviction of the scientist: this mobilization *must* join together a truth of more or less independent phenomena with the power of interpretation, and therefore must always be able to be extended further (the mentality of the *puzzle solver*). The scientist working under a paradigm cannot avoid being a "realist."

The question of progress had already changed meaning in the demarcationist tradition: from being the consequence of a healthy methodology, it

had become the condition, giving a de facto privilege to physics and the other experimental sciences in the strict sense. Here, the reversal of the terms is complete, for
the condition has lost all appearance of generality. The paradigm celebrates an event,
and it is this event that historians such as Hélène Metzger submit to, historians who
seek to reconstitute the interpretive ideas and systems of their actors. Suddenly, access is closed and, in order to discover the interpretive aspect, the solidarity with
the air of the time, it is now necessary to go through the scientists themselves,
through their work of reformulation, and no longer through their "context." For
language, here, loses its general power of interpretation in order to enter into a relation of *risky* invention with things.

A paradigmatic theoretico-experimental science can be recognized
not only by the singularity of its mode of fabrication of facts but also by its preoccupation with the *artifact*. We could say that every fact is here an artifact, a "fact
of art," but precisely because it is essential to distinguish facts depending on whether
they refer to a form of general, unilateral power or to an event-power. The artifact
the experimenter fears is an observable fact that he is convinced has been *dictated by
experimental conditions*, which are then recognized not as conditions of the staging,
but as the conditions of production, creative of observable phenomena. The risk of
the artifact singularizes the paradigmatic sciences in relation to the set of other sciences in which phenomena are subjected to the practices of the laboratory. The first
celebrates a phenomenon that allows itself to be staged; the second uses the general
power to subject anything at all to an imperative of measurement and quantification.

What brings us to this putting in question of the notion of the
paradigm, which links it to the singularity of the theoretico-experimental sciences?
Very precisely, it is a first approach to what Popper put under the sign of emergence:
a description of the social organization of paradigmatic disciplines as a consequence
of what will henceforth be their point of reference. "Before" the event, in the "pre-
paradigmatic" stage, a scientific practice is, according to Kuhn, in a state of double
dependence: in relation to facts of all types, which lend themselves to all sorts of
discordant interpretations; in relation to a social and cultural environment that is
equally interested in the facts, proposing interpretations, questions, visions of the
world. The scientist, then, must try to cultivate the virtues of lucidity and a critical
spirit, which is the only way of assessing the difference between these multiple other
interpreters of the facts. After the event, the difference between these multiple others
is created by transforming the mode of production of facts. It is the event that takes
advantage of the communities in order to make itself close in on them and decree
their conditions of production (transmission of the paradigm). The relation of social

force—the scientific community, the sole judge of "good questions"—intensifies a relation of force irreducible to the social, at least in the purely human sense. We thus understand why the practitioners of the paradigmatic sciences recognize themselves so well in Kuhn's description. The psychosocial dimension does not disturb them, for it translates,[11] sanctions, and, as we will see later, amplifies an irreducible difference from social analysis.

But the problem returns, for one of the essential attributes of the paradigm, its rarity, seems to be contradicted by an attribute equally essential to science as a historical tradition, namely, the claim to constitute a general enterprise of the production of intelligibility. Philosophers of science, who have failed to specify the criterion that grounds this claim, did not invent it. The academic structure that divides what we are dealing with into territories bearing the name of a science is not the simple product of a philosophical error. The notion of the paradigm can thus lead in turn to a position of denunciation: all sciences that do not operate by means of a paradigm are only ideological claims. Moreover, this is not far from Kuhn's own position, though he does not denounce the unfortunate "pre-paradigmatic" human sciences but simply feels sorry for them. This is what the practitioners of the theoretico-experimental sciences, on the other hand, are most often disposed to admit.

In fact, Kuhn's historical description is not historical enough. It does not teach us to laugh but only to celebrate. Most notably, it confuses the celebration of the event, in the sense that it creates a before and an after, with a celebration of the type of "progress" that follows the event. It also confuses "crisis" and "revolution," and does not take into account the fact that if the crises are, to a certain degree, forced upon scientists, revolutions, for their part, are constructed by scientists. Not every crisis will be proclaimed to be "revolutionary"; on the contrary, certain crises will be narrated in a style that accentuates the continuity of development, and not the break. Finally, it confuses the construction of borders between the disciplinary domain and the "exterior" with a naturally autonomous development of the discipline, which the "exterior" should respect under pain of hindering the inventiveness of the sciences. Without the paradigm, certainly, scientists would be unable to assess the difference between "good" questions, those authorized by the paradigm, and the questions that interest their contemporaries. In this sense, the paradigm inspires in scientists a certain passion for anything that allows them to make this difference be recognized. But this in no way means that a science working under a paradigm "is" autonomous, in the sense that it could be separated from the rest of society by a kind of "informational closure,"[12] letting in material resources, but determined by the only landscape of puzzles it engenders through its own dynamic.

In all these cases, Thomas Kuhn's description thus accentuates the image of a science developing in the way a natural phenomenon does, with "normal" evolutions marked by crises—an image that seems to be, if not produced, at least stabilized by the rhetorical strategies of scientists. To describe the life of the sciences as a natural phenomenon is to say that that there is only one choice: either to hamper them or to give them the means to continue. But if the historian recognized that the proclamation of a revolution, as the claim of autonomy, is a strategic move, if he regained his freedom faced with scientists who are themselves more free than they make him think, what type of laughter would he learn: the laughter of irony or the laughter of humor?

II

PART

Construction

F   O   U   R

Irony and Humor

Constructing a Difference

WHAT SHOULD we retain of the approaches to science we have now marked out, if not that this singular enterprise seems devoted to pushing its interpreters' backs up against the wall? Either, like epistemological philosophers such as Thomas Kuhn and Karl Popper, they seek a means of ratifying the difference to which scientists lay claim, or else, like Feyerabend and most contemporary sociologists of science practicing the so-called strong program, they seek to deny it any "objective" import whatsoever.[1]

In both cases, the instruments and finalities vary. Karl Popper never admitted his proximity to Thomas Kuhn, though they both celebrated scientific practice as the product of a novelty that escapes human intentions and calculations, and irreversibly transforms them. In one sense, "normal" scientists, working within a paradigm, are indeed typical examples of the subjects of the "second world" as re-defined by an inhabitant of the "third world," to which their anticipations, hopes, and practices are subordinated. In accordance with the epistemological tradition, Popper wanted to make scientific practice coincide with the ideal of critical lucidity. To the great scandal of the Popperians,[2] Kuhn depicted a social organization of the sciences that gave the inhabitants of the third world a maximal power, since it made the subjects of the second world the vectors of a "way of posing problems" without "asking oneself questions." In the same way, in the sociology of the sciences, the finalities and emphases vary between Feyerabend and the partisans of the "strong program." Feyerabend denounces relations of force and trickery, while the sociologists understand themselves to be doing their job, just their job. They do not denounce illusion, because according to them every human activity tends to present itself in a mode that is unique to it, to give a biased image of itself. They "only" claim the ability to do with scientific practices what they do with other practices, namely, to stage the difference between these practices and the image they give of themselves.

As for myself, the singularity of the sciences that I am seeking to construct would be rejected by the sociologists in question because it takes the scientists' scandal seriously when it reduces their claims to objectivity to a "particular folklore," susceptible to the same type of analysis as the folklores of other human practices. I must emphasize here that my project does not thereby seek to ground a privilege for the sciences, which alone would escape sociological analysis. The same type of question should be posed with regard to other practices. We know that certain ethnologists, such as Jean Rouch, present their films to "expert" members of the groups filmed and accept the test constituted by their reactions and criticisms. The "Leibnizian constraint" not to "go against established sentiments" here becomes a vector of knowledge: it constitutes one of the constraints in which the pertinence of the interpretation is put at risk.

In order to stabilize the difference between the "sociological approach," in the sense illustrated by the strong program in the sociology of the sciences, and the approach I am trying to practice, I will have recourse to making a contrast between "sociology" and "politics." This contrast does not designate a stable difference between what are called "sociology" and the "political sciences." It is rather a matter of "creating" this difference so as to demonstrate a divergence in their interests. I want to show that the singularity of the sciences does not need to be denied in order to become discussable. In order to make scientists actors like any others in the life of the city (the "political" preoccupation), it is not necessary to describe their practice as "similar" to all other practices (the "sociological" preoccupation). The quotation marks (which I will omit in what follows) indicate that the differentiation is related to the difference I am creating, without any ambition to define the specter of the real practices.[3]

I will start with an apparently trivial contrast. There are rather few veritable "theories" in the political sciences, which are today instead engaged in historical studies or a labor of more or less speculative commentary, which always depend on the situations and stakes created by history. By contrast, sociology remains haunted by the model of the positive sciences, those that can lay claim to a stable object in relation to history, authorizing the scientist to define a priori the questions it is proper to ask of every society.

This contrast can be attenuated. The ideal of the positive sciences does not define all of sociology, and many sociologists actively take into account the irreducibly historical and political character of any definition of what a society "is." Some of them also take into account the fact that their own activity as sociologists actively contributes to this definition. The important point, from the viewpoint of the difference I am proposing, is that *today* no sociologist engaged in this type of practice is unaware that he is participating in a "reflective," "nonpositivist," or "nonobjectivist" sociology. In other words, the ideal of a sociology copied from the model of the positive sciences remains dominant enough that no sociologist can be unaware of it.

I have decided to exploit this contrast because it seems to me to be capable of conveying a difference that is less empirical. Of sociology, it is necessary to say that it is the science of the sociologists: "society" as such brings together multiple actors, but none of these actors, except the sociologists, have any particular interest in defining what a society "is." The situation is very different in the political field. Politics, in the *practical* sense, in the sense that we can today say that it is, or should be, "everybody's business," is certainly what specialists in political science seek to understand; but they are always preceded by practices that are explicitly affirmed as political practices. In other words, from my perspective, the position of the commentator "following" history — which is the position of the specialist in political science — is not a weakness, but the expression of the fact that this specialist is situated among other actors who are asking questions similar to his own; who ceaselessly invent the way in which references to legitimacy and authority are *discussed* and *decided*, as well as the *distribution* of rights and duties, and the *distinction* between those who have the right to speak and the others.

The primary advantage of deciding to accentuate the difference between sociology and "politics" is that it clarifies the disquietude of scientists faced with the idea of a "sociology of science." It is difficult to count on a butcher for the quality of the meat. It is difficult to reassure scientists, as practitioners of the positive sciences, about the sociologists' claim to be "doing their job, just their job." They are aware of the actively selective character that allows a science "to give itself an object." They fear that what interests them in their activity might be actively eliminated by the sociology of science, as an obstacle to its own definition of what a "social object" is. Does not the "strong program" of the sociology of science take on the principle of assimilating their "proofs" and "refutations" to simple effects of belief?

We are returning here to the mobilizing power of words that claim to judge or explain. Sociology, as I am here defining it, gives itself as its legitimate ideal the power to judge, to unveil "the same" beyond the differences that merely belong to the lived experience of the actors. What do the thoughts of scientists matter? What do their "myths" of truth and objectivity matter? The duty of the sociologist of science is to ignore these beliefs in order to unveil what scientists are participating in, whether they know it or not, and the type of enterprise that defines them, whether or not they believe themselves to be "autonomous" actors. From this viewpoint, methodological differences—notably, for example, those that opposed sociologists who start with the actors and those who start with structures—count less than the common ambition: to define the "social" object in general, and to use this definition to select the common traits beyond the differences, which will then be termed "empirical."

According to the "difference" between sociology and politics that I am proposing—which I realize is radically asymmetrical—the relative absence of theory on the material of the political sciences takes on a positive significance. The specialist in political science deals with a dimension of human societies that is not the material for an "objective" definition, practiced "in the name of science," because in itself this dimension corresponds to an invention of definitions: Who is a citizen? What are his or her rights and duties? Where does the private end? Where does the public begin? These are modern questions, to be sure. But the fact that we recognize how the problems we are posing are expressed and regulated in other societies does not give the specialist the power of judging, but only the possibility of *following* the construction of the solutions that every collectivity brings to the problem.[4]

In one sense, Feyerabend's denunciation of the privileges claimed by Western science is indeed political, but in the sense that, instead of *following* the construction of this claim, it contests it. Feyerabend does not practice a political approach to the sciences, he *does politics*. The disappointment experienced by epistemology faced with the impossibility of grounding the legitimacy of science—and, of course, with the spectacle of the ravages committed "in the name of science"—has made the role of the analyst swing over to that of the actor. The aim of the "political" approach I would like to attempt is not to forbid this role changing but to clarify it. Political engagement is a choice, and not the result of a disappointment linked to the discovery of the political dimension of the practices that reason was supposed to regulate.

Great Divisions

Among the formulations, definitions, and inventions of the political, there is one that stands out insofar as it implies an explanation of the problem as such. "Politics" comes from a Greek word, but—and I am here referring to Jean-Pierre Vernant—the Greek city is less the admirable site where "our" democratic ideal was invented than the site where the various means through which a human society *constitutes itself* were put into words and problematized. Through what type of order, through what "arrangement" among those who are recognized as actors (in this case, this would be male citizens, not women or slaves) will political power construct itself? To this desacralization, which deprives power of the power to justify itself, there corresponds the Aristotelian definition of man as the "political animal."

Aristotle also happened to define man as the "rational animal." The tension between these two definitions is highly significant for our purposes. If it is "reason" or the "logos" that dominates, then politics will itself be subordinated and judged by the quality of its relationships with a nonpolitical authority, the Good or the True, which allows discordant and uncertain opinions to be silenced. The Sophists, experts in the logos that reorients, arranges, and creates opinion, must be condemned. This was Plato's position, this is the reading Heidegger gives of Aristotle, this is also the "established sentiment" that presides over the modern definition of a science "outside politics," which can only apprehend the possible play of politics in its midst in terms of impurity, lack, distance from the ideal. But what happens if, like Hannah Arendt, one puts in question this opposition between the (false) truth of the Sophists, for whom man is the measure, and rational truth, if one admits as one's starting point that "speech is what makes man a political being."[5] Once again we find ourselves in a situation of "irreduction," in which the modes "opinion" and "reason" lose their power of self-definition and are opposed to each other. It is then necessary to follow the way opinion and reason are defined in relation to each other, and in particular the type of test that presides over their differentiation.

It will be noted that this mutual definition concerns both politics and knowledge, which find themselves not confused but associated by the same type of problematization. The same question presents itself with regard to the person who claims to speak for others as it does with regard to the theory that claims to represent the facts: "How does one recognize the legitimate claimant?" We can, in this sense, speak of the birth of a politics of knowledge and of a science of politics at one and the same time. The solutions produced will be capable of diverging, and of selecting eminently different criteria; but it will always be a question of "arranging" and distributing, of defining rights and prescribing duties. The fact that, since Aristotle, politics had been traditionally defined by the concern to organize the common life of humans (*praxis*), whereas that which addresses itself to things (*poiesis*) was derived from an activity defined by utilitarian ends, was from this perspective part of a particular set of solutions, and not part of the problem. The stability of this solution depends on the claims, rights, and duties that the relation to things can or cannot give rise to.

From this perspective, the double definition of the political and the rational by the Greeks is new in that it *explains* the double problem of the legitimacy of power and the legitimacy of knowledge. The multiple and controversial solutions given to these problems do not divide up human history into those who were ignorant of politics and reason and those who "discovered" the problem, but they signal a difference whose consequences must be followed: claims to power and knowledge will now have to give an account of themselves. For the specialist in politics, the politologist, politics is not born with the Greek city, but the Greek city forces the politologist to recognize that its actors will henceforth be explicitly asking themselves questions similar to his own.

Rather curiously, an analogous problem arises with regard to the second "great division" that haunts our modernity. We are referring to the Greeks for the definition of reason we are putting to work, we who have invented the sciences, whereas all other human societies let themselves be defined by their tradition. We are referring to human traditions for the definition of "culture," we "humans" who are beings of culture, whereas all other "animal societies" let themselves be defined by the specific codes to which they are subjected. In fact, from the modern perspective, the two questions are one. As if the definition of the human as opposed to the animal found its full actualization with "us," we moderns who know ourselves, according to certain authors, to be "free," according to others, "rational." But the two criteria converge in that they are both opposed, through different aesthetics, to the same "illusions" of belonging and determination. Now the questioning of the "great division" between opinion and reason produced by Aristotle's "political" reading finds its analogue in the questioning of the great division between the human and the animal.

The privileged site where the division between man and animal is discussed is, of course, primatology. Classical primatology adhered to the thesis of the great division because it gave itself the mission of identifying the rules that the specific organization of a group of primates (for example, chimpanzees or baboons) would obey. In this sense, primate society was the dream of the "sociologist," as I have defined him: an object whose stability is guaranteed by the identity of the species, to which the individuals as well as their relations are submitted. Now, some contemporary primatologists have proposed a very interesting "heresy." After living among them, Shirley Strum concluded that baboons are "socially overendowed."[6] It seemed to her that the baboons she observed, by their very activity, were ceaselessly *creating* responses to the questions that classical primatology was asking its subject: What are allies? How does one make allies? Through whom does one have to pass in order to be accepted? Whom should one distrust? They would ceaselessly negotiate or renegotiate their roles, their mutual relations, their networks of alliance, the tests that identify a weak ally, or put one in question — in short, the very structure of their society. In other words, primatology must abandon the search for invariants to which individuals submit as members of a society in order to follow the construction of a social link insofar as it is, for the primates-actors, a problem and not a given.

It will be noted that I am here following a strategy of the "Popperian" type, in the sense that Popper characterized the three worlds in terms of the different questions that they *force* one to ask. Of course, the baboons did not address themselves to Shirley Strum to ask her to recognize their political behavior, nor were they scandalized to see it rejected by classical primatologists.[7] Nonetheless, Strum's narrative stages a quest for pertinence, at the end of which she must, since she defines herself as a scientist, assert that her study of baboons compels her to declare that her observations are incompatible with the idea of a submission to rules inscribed in the species.

If the baboons "do politics," in the sense that they ceaselessly *constitute* their societies, could the same be true for ants or rats? "Where should we place with certainty the beginnings of political behavior? Should we exclude social insects under the pretext that the major negotiations take place before the appearance of phenotypes?"[8] To this perplexing question, a single response is stable: the one that deals with the question of the words that the object we are dealing with forces us to use. To this day, it is the primates who have been able to make their specialists *explicitly* recognize in them a "speculative" activity, individual strategies that actively take into account an abstract notion of society that must be created or maintained. In this sense, the "politologist" of primates is hardly distinguishable from the "ethnomethodologist," for whom it is relations between actors that ceaselessly construct society, except that here it is not a question of "methodology." To this day, only humans have been able to impose a state of permanent controversy on their specialists with regard the question of knowing what comes first, actors or structures. For it is they who have imposed on themselves "heavy" differentiations, like those that *explicitly* disqualify certain social actors as political actors (women, slaves, and foreigners with the Greeks, immigrant workers and minors in France).[9]

The Political Invention of the Sciences

We are, from all appearances, very far from the question of the sciences. Yet are we as distant as we think? Whether it is a question of the indignation of scientists faced with the idea that their activity can be reduced to an object of sociology, or the question of the differentiation between those who have a right to intervene in a scientific debate and those who must be excluded from it, the question obviously being asked is that of the distinction between science and opinion. What is at stake in every question concerning the autonomy of the sciences is the distinction between those who have the right to intervene in scientific debates (the right to propose criteria, priorities, and questions) and those who do not have this right. The opposition of scientists to any sociology of the sciences can then be understood in *political* terms. The singularity of the primates is conveyed, as we have seen, by the fact that they have been able to impose on primatologists the nonpertinence of a gaze that would submit them to codes and rules from which their behaviors could be deduced; the singularity of scientific communities is conveyed by the fact that they demand of their environment that it recognize the distinction between the products of their activity and all other human productions.

Just as human politics is not reducible to the politics of baboons, the "politics of reason" I am trying to characterize is not reducible to the games of power we today associate with "political politics." To recognize a political dimension *constitutive* of the sciences is first of all to understand why the conflict between the sciences and their interpreters is a foreseeable one once the latter undertake to judge — that is, to relativize — the distinction between science and nonscience. Scientists, in the course of their history, have shown themselves to be remarkably tolerant, and indeed indifferent, to the means utilized by their interpreters to give an account of this distinction. On this subject, they themselves have advanced all sorts of interpretations, from pure positivism to a mystical quest. Putting the distinction in question, by contrast, is not a matter of interpretation but the subject of conflict. Whence the interest of a *political* approach to this distinction, an approach that allows a problematic space to be created where one will be able to *attend to* the construction of the difference between science and nonscience, in the same way that the politologist can attend to the consequences, on political life, of the Greek invention of politics as a problem.

Designating a problematic landscape in no way authorizes one to reduce the solutions that are inscribed in it to a common trait. The possible common traits, or the relations of resemblance, are derived from a comparison of the solutions, and not from an identification of the problem through these solutions. This also means that the analysis of the tests through which solutions of the political type are invented (Who are legitimate actors? How are propositions deserving of authority to be selected?) confers on the analyst no a priori superiority, no stable position of judgment. Analysts can submit themselves to a "principle of asymmetry," but only in the sense that asymmetry is a requirement they turn against themselves, a test they impose on themselves in order to try to escape the judgments of history they have inherited. But not in the sense that it would confer on them a right to judge, to lead differences back to a "same" shared equally by all solutions. The multiplicity, as a multiplicity of invented solutions, gives no superiority as such to the person who deciphers it. Rather, it institutes a relation of proximity with those who, since they do not share the tests we have invented for ourselves, appear to us, us moderns, so easy to disqualify. We here join up with the trajectory of *We Have Never Been Modern*, thanks to which Bruno Latour—in a difficult success—can posit, as the horizon of the new tests we will have to invent, the fact that "we are not all that far from the premoderns."

This is why, moreover, the history of the sciences constitutes the test par excellence for historical practices. For historians are also tempted to think of themselves as "modern," heirs of the great political division between scientific practice and opinion. For example, in order to historicize the passage from the epoch where "we did not yet know" the Earth revolves around the Sun to the one where "we do know," the historian might think that a "modest" solution is sufficient, a solution that consists of complicating the usual narrative by showing that the "discovery" does not have limpid simplicity attributed to it. But it is not enough to stop there, for historians do not suspend the certainties they themselves share with their contemporaries: the Earth is indeed a planet. What happened to our human histories when the Sun entered with them into this new relation that now forbids us from doubting that it is the Earth and not the Sun that "revolves"? For are they not, as historians, themselves the heirs of numerous social, political, ethical, affective, and aesthetic transformations to which we have all been subjected, whether or not we are scientists, and which, on balance, permit us to say, "You would have to be a crazy, dramatically ignorant, lunatic, or culturally backward person to doubt the movement of the Earth"?

This is why Bruno Latour can make the social history of the construction of scientific knowledges the focus of his argument that "we have never been modern." This implies, correlatively, that the only person who could do this history is a historian who would know what it meant for him "to have been modern," without for all that *denouncing* what he had been, or unveiling the trickeries and illusions of which he had been the victim; that is, without opposing the truths constructed by the sciences to another truth with a greater power—even if it is the a priori challenge of any truth that is not reduced to a belief "like the others."

I will call "humor" the capacity to recognize oneself as a product of the history whose construction one is trying to follow—and this in a sense in which humor is first of all distinguished from irony.

As Steve Woolgar has shown, the sociological reading of the sciences of the relativist type puts its specialists in the position of being "ironists."[10] They are those who will not let themselves count, who will bring to light the claims of the sciences. They know they will always encounter the same difference in point of view between themselves and scientists, which guarantees that they have conquered, once and for all, the means for listening to scientists without letting themselves be impressed by them. Some authors can advocate an "ironic" reading of their own texts because the latter are equally scientific (dynamic irony). The fact remains that the position in principle requires a reference by the author to a transcendence (stable or dynamic), to a more lucid and more universal power to judge that assures his or her difference from those being studied.

Humor, by contrast, is an art of immanence. The difference between science and nonscience cannot be judged in the name of a transcendence, in relation to which we would designate ourselves as free, and where only those who remain indifferent to it are free. For our dependence on this transcendence in no way reduces our degrees of liberty, our choice as to the way we will attend to the problems created by the constitution of this difference. The situation is the same as that of politologists, who know that their problem would have no meaning had not the Greeks invented an "art of politics." They are themselves a product of this invention, which they thus cannot reduce to nothingness. But they remain free to put this invention in history.

In this sense, irony and humor constitute two distinct political projects, two ways of discussing the sciences and of producing debate with scientists. Irony opposes power to power. Humor produces (to the degree it itself manages to be produced) the possibility of a shared perplexity, which effectively turns those it brings together into equals. To these two projects, there correspond two distinct

versions of the principle of symmetry: an instrument of reduction or a vector of uncertainty.

On the Event

There is a beautiful Talmudic narrative that puts on stage three rabbis confronted with
a point of interpretation of the Law.[11] Rabbi Eliezer, to make his viewpoint prevail,
has recourse to miracles: a carob tree is torn from the earth, a river starts flowing
backward, the walls of the house of study lean inward—but none of these arguments is judged to be admissible. Then Rabbi Eliezer appeals to the Most High,
and a celestial voice confirms his authority. But Rabbi Joshua rises and cites Deuteron-
omy: the Torah "is not in the heavens." The Most High has given the text to men so
that they can discuss it. He no longer needs to intervene in the discussion of the
text's meaning.

The scansion, the event that constitutes the gift of the divine
text, establishes a difference between before and after—but what is this difference?
What does this difference bear upon? When and how? The event does not say, and
the Jewish traditions tell us that this is the way it must be. A great number of actors,
all of whom have been, in one way or another, produced by the text, undertake to
draw lessons from it. All are situated in the space it has opened; none can claim to
have a privileged relation of truth with it.

The notion of event that I have just introduced allows me to
specify the relative positions of scientists and their interpreters. The decisive point
here is no longer to deny the differences scientists claim for themselves, but to avoid
any way of describing them which implies that scientists have a privileged knowledge of what this difference that singularizes them *signifies*.

It is the event that opens up this perspective—as long as one af-
firms that, as the creator of difference, the event is not for all that the bearer of signification. The invention of the "art of politics" by the Greeks was an event, it created a difference, but the signification this difference will take on, the solutions that
will be brought to the open problem, and the commentaries and criticisms these solutions will provoke are all part of what follows the event, not its attributes. The event
is not identified with the significations that those who follow will create for it, and
it does not even designate a priori those for whom it will make a difference. It has
neither a privileged representative nor legitimate scope. The scope of the event is
part of its effects, of the problem posed in the future it creates. Its measure is the
object of multiple interpretations, but it can also be measured by the very multiplicity
of these interpretations: all those who, in one way or another, refer to it or invent a

way of using it to construct their own position, become part of the event's effects. In other words, every reading — even a reading that denounces the event and calls it a fake — still situates the one who proposes the reading as an heir, as belonging to the future whose creation the event contributed to. In itself, no reading can claim to "prove" that in fact nothing in particular has happened. Only indifference "proves" the limits of the scope of the event.

Just as the event, in itself, does not have the power to dictate how it will be narrated or the consequences that will be authorized on its behalf, neither does it have the power to select among its narrators. Some of these narrators will try to augment to the maximum the scope and rights authorized by the event, while others will aim at minimizing them. Whoever undertakes this work will have, as his sole constraint, the recognition that he himself is the heir of what has taken place, that the event situates him, whether he likes it or not (cf. the retaliation to which relativists are exposed in matters relating to the sciences when they want an X-ray exam or a prescription of antibiotics), that is, the recognition that he himself is a constructor of the history that follows the event, one constructor of signification among others.

This indeterminate character of the event gives its meaning to the difference between philosophers and scientists, which is what concerns us here, given Thomas Kuhn's description. The scientists have recognized one part of the event, and have recognized themselves, practitioners of a normal science "provoked by the event." The philosophers, by contrast, were asking more: they were demanding that the history provoked by the event be capable of grounding its legitimacy. We here meet up with the contrast between the "foundation" [fondation] and the "ground" [fondement] proposed by Gilles Deleuze: "The foundation concerns the soil: it shows how something is established on this soil, how it occupies and possesses it; whereas the ground comes rather from the sky, it goes from the summit to its foundations, and measures the possessor and the soil against one another according to a title of ownership."[12]

The ironic relativist ceaselessly repeats and celebrates the failure of the philosophies of the ground. No title of ownership can measure the rights of scientists to possess the "soil" they occupy. They are convinced, to their own satisfaction, that no procedure recognized as scientific is capable of dictating, in controversial cases, the outcome that the "true scientist" would have to choose. According to the viewpoint I am defending, the scope of the demonstration is zero, for it assumes that the foundational event can give an account of itself. What scientists know, as I am trying to singularize them — thus excluding the systematic producers of artifacts "in the name of science" or "in the name of objectivity"—what their tradition tells them, is that the foundation has already given way to diverse reprises, that the soils have been occupied, that is, that the event can be repeated. No procedure, however rational it might be, and no submission to criteria, whatever it may be, can guarantee this repetition. But the repetition would not find the terrain where it could be produced were not the scientists acting with a view toward its production.

      If we can risk a parallel with the theory of grace (an interesting theory of the event), I would situate the position of scientists outside both the harsh perspective of Paul and Augustine, where God alone decides, whatever the actions, wills, and works of humans, and the soft semi-Pelagian perspective, according to which grace inevitably responds to the soul's movement toward God (which allows one to affirm that, even if man is incapable of attaining salvation without grace, an initial movement, of which he is capable, is enough to open up the path of salvation to him). Rather, they are situated in the perspective invented by Leibniz's monadology. No finite being has the power to know how to act, uncertainty reigns, with no way out; but we know that, in one way or another, this world is the best possible world; the only coherent attitude is thus to try to live in harmony with the principle of God's choice of this world, to seek to do the best one is capable of while hoping that the accomplishment of this best is included in the divine definition of the world. To the idea of the best of all possible worlds there corresponds the idea of propositions whose scientific character would be decidable. With neither guarantee nor promise of success. But not without precedent.

      Obviously, we still need to understand the type of events that create a precedent for scientists, and to understand them in a way that allows us to follow the construction of the sciences without either ratifying or denouncing them, to appreciate the engagement and passion of scientists without losing the possibility of laughing at them. With humor or irony, depending on the way they situate themselves in the scientific tradition: depending on whether they invent the means to prolong it, or claim it as their own in order to disqualify obstacles to its prolongation.

F   I   V   E

Science under the
Sign of the Event

In Search of a Recommencement

TO PUT the question of the sciences under the sign of the event is to accept—against the ahistorical criterion of rationality—the possibility of establishing a parallel with the way Gilles Deleuze and Félix Guattari characterize philosophy as a *contingent process*.

Philosophy was born in Greece. Should the power to explain this fact be attributed to the historical Greek singularity? Or, on the contrary, should this singularity be referred back to the general conditions that allowed thought to discover itself, conditions for a nonevent, for the passage to the reality of a possibility whose rights and duties would be derived only from itself? In *What Is Philosophy?* Deleuze and Guattari respond that Greek philosophy was not the "friend" of the city any more than modern philosophy is the friend of capitalism, but that neither the city nor capitalism are "neutral" milieus for a philosophy that would derive its right to exist from a universal, ahistorical imperative. The philosopher, in the Greek city, pushes to the absolute the problem of a community of men who will themselves to be free and to be rivals. How can we recognize the true friend of thought or the concept? How can we differentiate it from its simulating rivals? What tests should its statements be submitted to in order to distinguish them from opinion? How do these tests convey the power possessed by the concept to affirm its difference from opinion? The life of the city provides much more than a context for all these questions — the questions of Platonic philosophy — for they would have had no meaning elsewhere or beforehand. But they nonetheless create an event: against the solutions invented by the city for other problems, they set the demands of a problem these solutions neither imposed nor foresaw, but of which they constituted the terrain of invention.

The idea of a contingent process excludes explanation, which would transform the description into a deduction. It also excludes arbitrariness, which would insist on the contingency only in order to affirm, in a monotonous manner, that nothing has taken place, that the constructed significations and engendered problems are all valid because they are all relative to their context. The contingent process invites us to "follow" it, each effect being both a prolongation and a reinvention. "The contingent recommencement of a same contingent process, in different conditions."[1]

How, then, should we characterize the history of the modern sciences as a contingent process? It is not enough to speak, with Kuhn, of the contingent existence of those societies that have admitted or respected the autonomy of scientific communities. Nor is it enough to locate, with Kuhn, the contingent advent of a paradigm. In both cases, as soon as it finds the occasion of its beginning, the contingency would preside over the advent of a process endowed with its own necessity. In order to avoid simply ratifying what is, I will have to try to interpret the ensemble of modern sciences, those that are and those that might be, that is, to prolong, to reinvent, "to recommence with other givens." This is why I have to invent a new mode of astonishment, a point of interrogation that does not commit me to privileging the experimental sciences, and to identify a "motif" (in the double sense, both musical and desiring) that would singularize "science" and make it capable of becoming, certainly not an object of definition, but a subject of history.

My astonishment, like my motif, is going to take me back to Galileo. In the wake of so many others, for Galileo's scientific work — but also the "Galileo affair," his condemnation by the church — constitutes a quasi-obligatory reference for all narrations of the origin of modern science. And this reference is not merely a historical artifact. Galileo himself appears to have been perfectly conscious of the fact that, with him, something new was in the process of coming into being. His public work celebrates an event — not only a "new world system," but also a new way of arguing, to which he attributes the power to make his adversaries give way through ridicule, and to force Rome to bow down and modify its interpretation of the Scriptures. In other words, Galileo presents us with both the problem of an event and the first explorations of its effects, and the signification that Galileo himself confers on it insofar as he is created-situated-produced by the event.

What subject of astonishment emerges with Galileo? I would like to situate him "before" the astronomical controversy, and thus before the Galileo "affair" properly speaking. As a first approximation, in any case, I take Galileo the astronomer to be inscribed in a history that he does not invent. Certainly, the telescope allows him to make observations that were inaccessible to others, and thus to present original arguments. But it is enough to listen to the anxious tones of Kepler, who is asking for a telescope, who would give his soul for a telescope, to conclude that, despite the controveries he aroused, Galileo's use of the telescope is not enough to singularize him. Galileo's astronomical work can, without too much difficulty, be judged by historians, who will pose the problem of his refusals—of Kepler's ellipses, for example—and will admire the formidable intelligence of his arguments. By contrast, historians hesitate when faced with the work of Galileo as the creator of the mathematical description of the accelerated motion of heavy bodies. How can they recount the production of something that, in its essentials, is still accepted by physicists, and that is still taught in the schools? How can they locate in history something that, since Galileo's time, seems to have resisted history? How can they explain that, when we see an inclined plane, we are always the near contemporaries of Galileo?

This would be my subject of astonishment: this force of an oeuvre that has remained stable, capable of triumphing over the relativity of opinions and viewpoints. This was a subject of astonishment for many philosophers, beginning with Kant, when they realized the scope of what that science, which debuts with Galileo, implies and imposes: a new type of truth. But Kant's example precisely warned of the dangers of this astonishment, of the slippery slope it involves. For the Kantian question—how to retranslate in an admissible philosophical mode the fact that Galileo (and Newton) indeed seemed to have made nature speak, to have made it confess its laws—manifests an astonishing *disproportion* with what Galileo in fact did, namely, describe a motion whose prototype is the descent of highly polished balls down the length of a smooth inclined plane, or the eternal oscillation of an ideal pendulum.

My subject of astonishment would then be displaced slightly: How can we understand, whatever the interest of rolling balls or an oscillating pendulum, the fact that we, who are, like Kant, the heirs of the event of their description, are so easily led to describe it as "the discovery of the laws of motion," and not, for example, as "the practical identification of the (limited) class of accelerated motions whose prototype is pendular motion or the fall of bodies in the absence of friction"?

We now come to the motif that, to me, seems to singularize the modern sciences as such. If normative epistemology failed to identify a criterion of demarcation between science and nonscience, we need to recognize that the search for such a criterion could seem to be justified. Ever since Galileo constituted the reference for what we now call "modern science"—a power to which another power, that of the church, would eventually give way—the question "Is it scientific?" has become the decisive question, the question that arouses passions and provokes invention, the question on which the raison d'être of the sciences apparently depends. This question is not identical with the question concerning the validity or falseness of a proposition; it precedes it, something that Popper had indeed seen from the start, when he refused to identify the scientific proposition with the valid proposition.

But do the norms that seemed to be evoked by the question "Is it scientific?"—if they cannot be identified by the epistemologist-judge—merely amount to simple affirmations that the ironic sociologist would be free to interpret, that is, to reduce to "a repertory of discourses available to justify actions undertook for completely different reasons"?[2] In other words, did Galileo "fabricate" the reference to science in his attempt to vanquish the power of Rome? Or were *Galileo and his struggle against Rome provoked by the event that constitutes the possibility of affirming "This is scientific!"*? It is this second viewpoint I will try to adopt. According to this viewpoint, what singularizes science is not the submission to criteria that would define a scientific procedure. The common "motif," taken up in different practical modes and regimes, repeats the invention that makes the response to the question "Is it scientific?" decidable—at a given moment and in a given domain.

Obviously, we are not yet done with the ironist, who of course will be able to point to a remarkable tautology: what is scientific is whatever scientists, at a given moment, decide it is. The position of the humorist, which I am trying to make my own, takes into account the passion, the relentless effort, the risk. If the response to the question "Is it scientific?" is a construction of scientists, it is not the fruit of an agreement among scientists, deciding among themselves something a detached observer can recognize as always undecidable. The gaze that sees the same, the undecidable, where those he is observing have as their raison d'être to create difference, is the gaze of power.

In fact, as I am now going to show, relativist skepticism—which reduces the difference that the scientist claims to create to the same, to the undecidable—is nothing new. It even constitutes, one might say, the "primal scene," out of which is born the singularity of what we call "the modern sciences."

The Power of Fiction

It is in the course of the third day of the *Discourse concerning Two New Sciences* that Galileo, under the mask of his spokesman Salviati, states the definition of uniformly accelerated motion, and I would like to understand how and why this "creates an event": "By steady or uniform motion, I mean one in which the distances traversed by the moving particle during any equal intervals of time, are themselves equal."[3] It is not without interest to see how Galileo himself will stage the event, that is, how the interlocutors Galileo has provided for Salviati—Sagredo and Simplicio—are going to react. The question is all the more interesting in that the roles of Sagredo and Simplicio changed between the *Dialogue*, written in 1633, and the *Discourse*, which was composed after his condemnation, in 1637.

In the *Dialogue*, Simplicio represents all of Galileo's adversaries, whereas Sagredo is the man of good sense, the man with whom Galileo's readers should identify. This strategy, moreover, has a formidable efficaciousness, for when Sagredo, forgetting his supposed impartiality, allies himself with Salviati in order to shower insults on the unfortunate Simplicio, and with him all those he represents, it is we readers who are, at the same time, made to participate in a veritable intellectual lynching. The new type of truth invented by Galileo is openly announced in the *Dialogue* like a truth of combat, verifying itself by its ability to silence or ridicule those who contest it. But in my reading hypothesis, which privileges the science of motion over the astronomical controversy, it is also announced in a quasi-clandestine manner. The composition of the *Dialogue* concentrates one's attention on the astronomical debate, and it is in its service—notably to show that the Earth can be in motion without our being aware of it—that the statements on movement are presented.

In the *Discourse*, the tone changed. Galileo has been condemned. Now an old man, he knows his death is at hand. He is writing clandestinely for readers he will never know. He is writing for the future, more for his successors than for the public. Theorems, propositions, and corollaries are lined up in good order. Simplicio and Sagredo have become simple stand-ins, asking the questions and posing the objections Galileo needs in order to bring to light the novelty and signification of his proposals.

When Galileo states his definition of uniformly accelerated motion, it is Sagredo who reacts:

> Although I can offer no rational objection to this or indeed to any other definition, devised by any author whomsoever, since all definitions are arbitrary, I may nevertheless without offense be allowed to doubt whether such a definition as the above, established in an abstract manner, corresponds to and describes that kind of accelerated motion which we meet in nature in the case of freely falling bodies.[4]

It thus seems that Galileo is expecting that the principal misunderstanding, the one he himself is the first to raise, will be the result of a *skeptical* reaction. His statement could be confused with an abstract definition, which refers to an author in the sense that this author, whoever he might be — there is no reason to take offense at this — does not have the power to cross the distance between the abstraction he creates and the world where, notably, bodies fall naturally.

In other words, Sagredo is a "relativist" *avant la lettre*: no author of an abstract proposition has the means to make nature a witness in order to carry the decision concerning its truth. The rivalry of human, purely human, points of view is unsurpassable. Every definition is arbitrary. Every definition, we shall say, is a *fiction*, referring to an author.

What authorizes us to make this observation? Nothing, if it were a matter of constructing a historical thesis. Slightly more, if we recall that Sagredo is not an author but a fictional character, and thus he conveys the diagnostic posed by Galileo himself, not in a "neutral" situation, but on the point of optimal encounter between the force and novelty of his exposé and the reactions of the educated public, the "savants" to whom he is addressing himself. In the *Dialogue*, Sagredo never hesitates to draw the most realist of conclusions from the astronomical demonstrations of Salviati, who never ceases to call him back to prudence. Galileo was thus able to plead that he himself (Salviati) would not encourage but rather discourage such excesses, contrary to the decision of Rome. It was not his fault if the "public," represented by Sagredo, refused to understand. In the *Discourse*, where it is a question of science and not the system of the world, Galileo thus seems to anticipate a rather different reaction from the public he is trying to interest. He has to impose himself "despite" the relativist skepticism that will greet, he fears, any abstract proposition, whoever its author might be.

The "relativist" reaction that Galileo stages is not without analogy to the argument that the Roman power had opposed to his own claims. Cardinal Oregio, who had become the personal theologian of Pope Urban VIII, has left us his recollection of the interview that the latter, then Cardinal Maffeo Barberini, had with Galileo after the first condemnation of 1616.

> He asked him whether it was beyond God's power and wisdom to arrange and move the orbs and the stars in a different way while yet saving all the phenomena displayed in the heavens, all that is taught about the stellar motions— their order, position, relative distances, and arrangement. If you want to maintain that God cannot and knows not how to do this, you must, added the prelate, demonstrate that all these things could not be obtained by a system different from the one you have conceived, that such a system would involve contradiction.[5]

The great scholar, concludes Cardinal Oregio, remained silent.

The fact that Urban VIII, upon discovering his own argument in Simplicio's mouth at the end of the *Dialogue*, concluded that Galileo had thereby meant to ridicule him, since everything Simplicio says is by definition ridiculous, belongs to the legendary history of Galileo's condemnation, which I will not linger over here. On the other hand, the argument itself interests me because it breaks the staging that Galileo himself had elaborated, and which is all too often taken up by those seeking to characterize the singularity of the so-called modern sciences. Galileo's adversaries were not only the belated heirs of Aristotle, which would have the effect of putting the Middle Ages in parentheses. The truth announced by Galileo not only had to impose itself against another truth that it contradicted. It first, and above all, had to to impose itself against the idea that all general and "abstract" understanding is essentially a fiction, that is, that human reason does not have the power to link up with the reason of things, that the latter refers to the order of Aristotelian causalities or to mathematics.

We know that when Barberini, the future Urban VIII, evokes God's omnipotence ("God can do anything that does not imply a contradiction"), he is taking up the famous argument of Étienne Tempier, the bishop of Paris, who, in 1277, condemned the entirety of the cosmological theses derived from Aristotelian doctrine on this basis. In particular, he condemned the proposition according to which "God could not imprint on Heaven a movement of translation," because the demonstration of this proposition rested on the absurdity of the hypothesis of the void, the production of which would be implied by such a motion. Absurdity is not contradiction. What appears to be absurd to us is perhaps not so for God. The authority of the argument appeals to an absurdity, namely, the idea of a rationality that, in one way or another, could pride itself on the power to establish the difference between the possible and the impossible, the suitable and the unsuitable, the thinkable and the inconceivable. It is this power that would come to refute the reference to the omnipotence of the divine author of creation. If God had so willed, what seems normal to us would not be so to him, what seems inconceivable or miraculous to us would be the norm. God's omnipotence implies that we were thinking on the basis of a risk, that we were daring to think, for example — as Samuel Butler did in *Erewhon* — that there could have existed a society where illness and misfortune were severely punished, while crimes and misdemeanors evoked pity and the most attentive medical care.

If no other difference between the imaginative and fictive world and our world can be legitimately invoked except God's will alone, which has chosen to create the latter and not the others, then any mode of understanding that is not itself reducible to the pure observation of the facts, and to the logical reasoning derived from the observed facts (bringing into play the principle of noncontradiction that even God respects), is of the order of a *fiction*, more or less well constructed, "elaborated in the abstract." In other words, the logicist definition of science against which Popper fought, the one that understands a scientific proposition to be a proposition logically derivable from the facts, was, according to Tempier's prescriptions, nothing less than the only nonfictive form of understanding. Now, the group of authors we have examined, from Popper to Feyerabend, and from Lakatos to Kuhn, are in agreement on a single point: scientific practice does not conform to these prescriptions; no "fact" intervening in a scientific reasoning is "observable" in a neutral manner, and no scientific reasoning is reducible to a logical operation admissible by the "facts"; they all form a part of the "elaboration in the abstract."

What should we think of the apparently contemporary character of the debate we have uncovered at the origin of the modern sciences? First of all, it seems to me, it signals the fact that something happened between Antiquity and this origin, this modern origin. The Greeks, had they been confronted with the postulate of a divine omnipotence, defined as the absence of constraints, would undoubtedly have denounced the ugliness of the hubris, of the pride that exceeds all limits, of the despotic decision that draws its glory from its own arbitrariness. I will not discuss here the various ways that philosophers have tried to restore the virtues of wisdom to the despot God — and I am, of course, thinking primarily of Leibniz — nor the thorny question of knowing how to recount the history that produced this figure of power, in relation to which human reason is called upon to situate itself. For Pierre Duhem, the philosopher-physicist, it is the unique glory of Christianity to have created, against the certitudes of the tradition, a dramatic distance between necessary truths and truths of fact, which it is possible to deny without contradiction. For the philosopher Éric Alliez, this history is first of all that of the cities, where, at the end of the Middle Ages, the difference between the possible and the impossible was a question of will, speculation, and the entrepreneurial spirit, rebelling against everything that would, in principle, make what is and what must be coincide.[6] In a case like this one, moreover, there is probably no choice to be made. If the words and actors are authorized by the Christian faith, they do not tell us why they are seeking this particular authorization, or why they find it in faith.

Let me emphasize, however, that the statement of Bishop Tempier, who pronounces these words and actualizes this authority, concerns a *political* problematic: it is a question of administering the renaissance of the pagan "Greek heritage," that is, of deciding which parts of this heritage can be considered to be the production of a "naked reason," uncontaminated by paganism (in this case, it will be logic, that is, mathematics), and which parts must be considered as suspect, tainted by their pagan source — a problem that is not without analogy to the modern question of the relations between "pure" science and ideology.

In any case, we should not underemphasize the importance of this fact: the Middle Ages created a new figure of skepticism, a figure in which skepticism, which is probably present in all human civilizations, was no longer formulated by a minoritarian thought, accepting the risk of exclusion or marginality, but by a thought that established explicit links not only with power, but with *a repressive dimension* of power. This skepticism *disqualifies* anyone who does not submit to its negative norms and instead undermines their obviousness, at their own risk and peril — and it can do so because it is authorized by a constraint imposed by *power itself*, condemning as erroneous, from the viewpoint of faith, any use of reason that would limit God's absolute freedom. Correlatively, this type of thought imposes *the power of fiction* as an unsurpassable horizon of our arguments, a power that has the language to invent "rational arguments" to bend the facts, to create illusions of necessity, and to produce an apparent submission of the world to its definitions "elaborated in the abstract." Any definition or explanation that, by going beyond the facts and logic, can thereby be persuaded to encroach on God's full freedom has ceded to the power of fiction.

The fact that this power of fiction has become the principal weapon of contemporary relativists, that the positivist adulators of scientific rationality have tried to prove that this rationality was shielded from it, and that Sagredo himself had recourse to it indicates that the argument was able to acquire an autonomous plausibility, the now "exotic" reference to divine omnipotence no longer being necessary to sustain it. In the perspective I am constructing, it is the obviousness of this power of fiction that constitutes not only the "terrain of invention" for modern science, but also *the means by which it will stabilize itself so as to better detach itself from it.* In other words, the contingency of the origin—and we should recall that nominalist skepticism, of course, hardly defines medieval thought as a whole—does not here define an "occasion" that could then be forgotten, but is captured by the procedural logic that constitutes this origin as one of its conditions. Wherever a "new use of reason" is produced—and this is how I propose to identify the singularity of the modern sciences—it will imply and affirm the inability of reason alone to vanquish the power of fiction.

A New Use of Reason?

The staging I have just indulged in does not seek the title of a historical truth, but merely the construction of a viewpoint from which the modern sciences could be understood as a contingent process. The fact that Galileo, at the very moment he bequeaths to posterity the science of uniformly accelerated motion, deliberately makes reference to what I am calling the "power of fiction," is, for me, the sign of the event. The force and novelty of his statement lie in the fact that it *can short-circuit* the argument that is staging this power, that it can oppose to it a counterpower that silences the skeptics...including today's relativists. "To begin again with other givens."

Among these other givens, there first of all figures the new inseparability between science and fiction. No legitimate use of reason can any longer guarantee the difference between what it would authorize and what it would relegate to fiction. As opposed to the dominant modern philosophy, which seeks a philosophical "subject" capable of offering this guarantee—a purified subject, stripped of anything that would lead it to fiction—the positive sciences do not require their statements to have a different "essence" from creatures of fiction. They demand—and this is the "motif" of the sciences—that they be very particular fictions, capable of silencing those who claim "it's only a fiction." For me, this is the primary meaning of the affirmation "This is scientific." This is why the search for norms was in vain. The decision as to "what is scientific" indeed depends on a politics constitutive of the sciences, because what is at stake are the tests that qualify one statement among other statements—a claimant and its rivals. No statement draws its legitimacy from an epistemological right, which would play a role analogous to the divine right of traditional politics. They all belong to the order of the possible, and are only differentiated a posteriori, in accordance with a logic which is not that of judgment, the search for a ground, but that of the foundation: "Here, we can."

Read in this register, the Galilean event can also make sense of the astonishment that I have taken on as a challenge. For it would indeed be *a new "use of reason,"* capable of doing what it was no longer believed possible to do, celebrating the statements that lightheartedly cross the distance between "nature" and polished balls rushing down a smooth, inclined plane. What is presented as having been reconquered in principle, if not (still) in fact, is precisely *something one believed to have been lost: the power to make nature speak,* that is, the power of assessing the difference between "its" reasons and those of the fictions so easily created about it.

It remains to be seen what there is about Galileo's statement concerning falling bodies that must not be "only a fiction."

The response to this question has often been given in a general mode. Thus, as everyone has said and repeated, Galileo's science of motion would be new in that it does not say *why* heavy bodies fall the way they do, but merely specifies *how* they fall. This distinction is always present today. When Stephen Hawking envisions the "end of physics," the construction of an equation that will tell us what the universe is, he hastens to stage the final act, where philosophers, scientists, and ordinary people will get together to discuss "why" the universe is as it is, and why we ourselves, who have identified it, exist. It is then and only then—when we have all reached an agreement on this subject—that we will finally know the thought of God.[7]

This example is enough to show that the question "how" cannot be identified with a humble prejudice, itself guaranteeing a difference between science and fiction. Rather, it entails a principle of distribution that decides who is entitled to speak. No matter how far he or she goes in inventing modalities of the question "how," the scientist is always working with other scientists. Galileo's statements have been subjected to different modifications, but their authors are scientists, they belong to the class of those who recognize themselves as Galileo's dependents. These modifications thus have the right to be qualified as "progress." By contrast, once it is a question of "why," the scientist admits that the stage becomes occupied by all those who had been excluded: the philosophers, and even ordinary people (if the former are admitted, how can the latter be excluded!). He will no longer claim any exclusivity, but he will claim, of course, that the "why," which is everybody's business, is the "why" whose "how" he has identified. According to Hawking, for example, when it is a question of the universe, the philosopher who thinks becoming or the event falls silent. The stage on which he will finally be able to speak will be defined by the equation that allows him to affirm that the universe IS.

The scientific "how" thus has no other a priori limits than those of the questions that, rightly or wrongly, are recognized as scientific. The "why," in this staging, has no autonomous formulation. It transcends the "how" only in appearance: it must first learn from the "how" what it is authorized to ask.

Thus, the differentiation between how and why is not a symmetrical division, but a distinction between a dynamic power, that of science, and its remainder, which is constantly reformulated as a repercussion. A game of fools, which was given its rules when Kant handed over to the power of science the whole of the phenomenal world, including the subject, insofar as it is "pathological," that is, explainable through reasons, motives, opinions, and passions — everything the "acting," "free," "intelligible" subject has to cut itself off from in order to determine what it *must* do.[8]

The new "use of reason" the Galilean event celebrates thus has two interesting features. It invents, with regard to things, a "how" that defines the "why" as its remainder. It selects those who can participate in the discussion of the "how," of its extension and modification, and defines the others, the philosophers and ordinary people, as those who come afterward — a landscape structured by a stabilized division between what is "scientific," the business of scientists, and the remainder. Both of these features are political. The first is addressed to things, and prescribes the way they should be treated. The second is addressed to humans, and distributes competencies and responsibilities in this treatment. Rome, Galileo proclaims, need not enter the territory of the sciences, which alone are fit to discuss whether the Earth or the Sun turns around the other. The "criterion of demarcation" that Popper's disciples vainly sought to define is thus indeed consubstantial with science. But it does not merit this because of a "rational" use of reason; it marks out territories invested *against the power of fiction* by those who are inscribed in the tradition inaugurated by Galileo.

But how does Galileo prove that his fiction is not a fiction like the others? What argument does he oppose to Sagredo's objection, who suspects that his definition of accelerated motion is arbitrary, like all definitions elaborated in the abstract? He takes the objection to heart, and even has Salviati say that he has discussed the problem with the author (Galileo). Then, he specifies what he means by "moments of speed." There is here a break in the style of Galileo's narrative, which must be confronted by historians who take him as their subject: there is the Galileo whose ideas on "motion" they try to reconstitute, and the Galileo who now tries to explain himself, and who apparently deems it suitable to paraphrase his theses, which correspond to our own. A Galileo who gives himself the luxury of playing the historian of his own ideas, of the difficulties he is experiencing here "at the beginning."[9] Galileo then establishes a differentiation between the *causes of acceleration* (the "why"), on which "different philosophers have expressed different opinions," "imaginings" whose examination would not be very "profitable," and the *properties of accelerated motion*, which he will demonstrate are indeed applicable — this is what is at stake — "to the registers animated by a naturally accelerated falling motion."

In other words, not only has Galileo staged the objection of Sagredo and the "power of fiction" it implies, but he calls on this power to disqualify what, in motion, is a matter of opinion and to indicate what is a matter for demonstration. Galileo's inquiry thus needs to affirm the power of fiction: it is that *against which* science must differentiate itself, and that *through which* it defines-disqualifies everything that is not science.

Then Galileo-author, that is, the trio thanks to which he is arguing, effaces himself. Theorems, corollaries, propositions, and problems will succeed each other. A succession on which very few relativist historians, such as Feyerabend, have dared to comment, but where the physicist is perfectly at ease: the difference is made, "his" Galileo is at work. "Reduce that to sociology," try to show in what and to what Galileo's response to this problem is relative, for example: "Given a vertical and a plane inclined to it, of the same height and having the same upper terminus; to find a point, vertically above the common point, from which a moveable object, falling and then deflected along the inclined plane, consumes the same time in this plane as [in fall] from rest through the [given] vertical" (Problem XII).[10] Galileo effaces himself in order to leave "speech" to the thing that will silence the others. *Enter the inclined plane.*

The Inclined Plane

According to Stilman Drake, Galileo became "our Galileo" in 1607.[11] In any case, it was in 1608 that there appeared, in Galileo's working notes, a schema over which historians would spill much ink. If, according to Drake, the author of this schema is "our" Galileo, for others, it describes his act of birth. By all accounts, it is a question of a "knot," an effectively realized experiment, and the person who performed it should have known or indeed already knew, or comprehended, "how" it was suitable for describing the motion of falling bodies.[12]

The schema that figures in folio 116v represents the distance between the point of impact on the ground and the edge of the table from which the balls fell—balls that, before rolling across the table, had (no doubt) descended the length of an inclined plane sitting on this table. In the calculations that appear on this page, Galileo correlates the distance to the ground with the vertical height the ball fell before rolling on the table.[13] By all accounts, the schema articulates three types of motion: the first motion of falling, which is simply characterized by the height of the fall; the horizontal motion on the table; and the motion of free fall, characterized by the horizontal distance it permits the ball to cross (for a table of doubled height).

This schema represents an experimental apparatus, in the modern sense of the term, an apparatus of which Galileo is the *author*, in the strong sense of the term, because it is a question of an artificial, premeditated setup that produces "facts of art"—artifacts in the positive sense. And the singularity of this apparatus, as we will see, is that it *allows its author to withdraw*, to let the motion *testify* in his place. It is the motion, staged by the apparatus, that will silence the other authors, who would like to understand it differently. The apparatus thus plays on a double register: it makes the phenomenon "speak" in order to "silence" the rivals.

What the phenomenon thus staged bears witness to is not trivial. The three types of motion it articulates are characterized by three different modes. The first fall permits one to characterize the moving body as having *gained* speed, and suggests that the speed gained is determined solely by the height of the fall. The horizontal motion is characterized as *uniform*, and the apparatus suggests that one attribute to it as speed (in the traditional sense of the relation between the distance traveled and the time it took to traverse it) the speed gained during the preceding fall. The third motion, that of the free fall, can only measure this speed if one admits that it is *composed* of two motions that do not interfere with each other, the accelerated motion of vertical fall, in a time that depends solely on the height of the table, and the uniform horizontal motion that is going on during the same time.

Not only does Galileo's apparatus articulate three different types of motion, it also presupposes and affirms the possibility of defining three distinct and articulated concepts of speed: speed in the sense that it is gained, linked to a past in which the moving body changed altitude; speed in the sense that the body "has" it at a given moment, and, for example, at the end of this fall, at the moment the body passes from the inclined plane to the horizontal table; and the speed of the motion insofar as it characterizes the horizontal, uniform motion of the moving body. The apparatus proposes an operational relation of equivalence between these three speeds: the *instantaneous* speed characterizing the moving body at the end of its fall is equal to that which it had gained *in the past* and it is also equal to that which *in the future* is going to characterize its uniform motion.

I have explained what Galileo's apparatus implies and affirms in order to show that the "law of motion" is not linked to observation but is relative to an order of created "fact," to an artifact of the laboratory. But this artifact has a singularity: the apparatus that creates it is also able, certainly not to explain why motion lets itself be characterized in this way, but to counter any other characterization. In effect, it can place the three motions that constitute it in variation: the height and slope of the inclined plane, the distance between the end of the plane and the edge of the table, the height of the table. To any disputation, then, a response can be invented (if the case arose, one could use two inclined planes, or make a comparison between a parabolic free fall and a vertical free fall).[14] The apparatus could then be seen as the generator of a set of cases, each responding to a possible putting in doubt, and in each case affirming that only Galileo's description is faithful to it. The different falling motions that can be observed have given way to a motion that is both unique and decomposable in terms of *independent variables*, controllable by the operator and capable of forcing the skeptic to admit that there is only one legitimate way to articulate them.

Obviously, nothing of all this figures in folio 116v, and Galileo invented other, much more picturesque stagings in the *Dialogue*. But the apparatus created in 1608 makes the world, which Galileo made his readers discover in terms of thought experiments, exist in the laboratory. We could certainly say that this is an abstract, idealized, geometric world. But we will have said nothing, for we will simply have repeated Sagredo's skeptical objection: this is simply a world answering to a definition elaborated in the abstract. The question is rather that of knowing what had been abstract, what singularizes this fiction. The fictive world proposed by Galileo is not simply the world that Galileo knows how to interrogate, it is a world *that no one could interrogate differently than he*. It is a world whose categories are *practical* because they are those of an experimental apparatus that he invented. It is in fact a concrete world in the sense that this world allows him to welcome the multitude of rival fictions about motions that compose it, and to make the difference between them, and designate the one that represents it in a legitimate manner.

Galileo's world appears as "abstract" because many things have been eliminated, whose categories the experimental apparatus does not permit to be defined. But the "abstraction" is here the creation of a concrete being, an intersecting of references, capable of silencing the rivals of the person who conceives of it.

Sagredo was not silenced because he was impressed by the subjective authority of Salviati, nor because he would have been led to recognize the well-foundedness of the proposed definition by some intersubjective practice of rational discussion. It was the experimental apparatus that silenced Sagredo, that forbade him to oppose another fiction to the one proposed by Salviati, because this was precisely its function: to silence all the other fictions. And if, after three and a half centuries, we are still teaching the laws of Galilean motion and the apparatuses that allow us to stage it—inclined planes and pendulums—it is because until now no other interpretation has succeeded in undoing the association, invented by Galileo, between the inclined plane and the behavior of falling bodies.

When we speak of "abstract scientific representation," we are too often referring to a general notion of abstraction, common, for example, to both physics and mathematics. But here, abstraction expresses an event and not a general procedure: the local, conditional, and selective triumph over skepticism. It was rather the medieval notion of speed that was abstract in the general sense, separable from the moving bodies it qualified: give me a means to measure space and time, and you will be able to forget the difference between the stone that falls, the bird that flies, or the horse that, exhausted and breathless, will soon collapse: I will tell you their speed, the relations between the space traversed and the time it took to traverse it. For Galileo, these movements are not all equivalent. His apparatus allows him to stage the movement of the stone, but not that of the bird. The speed of Galilean bodies—the speed that, we would say today, defines classical dynamics—is inseparable from the moving bodies it defines by the existence of an experimental apparatus, which permits one to hold, faced with the concrete multitude of rival propositions, that this speed is not merely one way among others of defining the behavior of this body.

Abstraction is not the product of an "abstract way of seeing things." It has nothing psychological or methodological about it. It is relative to the invention of an experimental practice that distinguishes it from one fiction among others while "creating" a fact that singularizes one class of phenomena among others. This is why the difference between what can be the "object of representation" and what is supposed to "escape" representation cannot be grounded a priori by a theory, philosophical or otherwise. To ground always means to refer to a criterion that claims to escape history in order to constitute a norm. Before Galileo, who would have held that Galileo's speed was "representable"—an instantaneous speed in which a body traverses no space at all in no time whatsoever? Who believes it is possible to "represent" light, which is neither a wave nor a particle, but which can, depending on the circumstances, respond to the representation sometimes as a wave and sometimes as a particle? The sciences do not depend on a possibility of representing something it would be the task of philosophy to ground; they invent possibilities of representing, of constituting a statement that nothing a priori distinguishes from a fiction, as the legitimate representation of a phenomenon. As Bruno Latour has emphasized, scientific "representation" here has a meaning closer to the one it has in politics than in the theory of knowledge.

S I X

Making History

Negative Truth

IN THE modern sciences, we can see the invention of an original practice of attribution of the quality of being an author, playing on two senses that it opposes: the author as an individual, animated with intentions, projects, ambitions; and the author as someone who creates authority. It is not a question of a naïveté, which could be critiqued, for instance, by contemporary theories of literature, but of a rule of the game and an imperative of invention. Scientists know themselves and their colleagues as "authors" in the first sense of the term. This matters little. What matters is that their colleagues are constrained to recognize that they cannot turn the quality of authors into an argument against them, that they cannot localize the flaw that would allow them to affirm that someone who claims to have "made nature speak" has in fact spoken in its place. This is the very meaning of the event that constitutes the experimental invention: *the invention of the power to confer on things the power of conferring on the experimenter the power to speak in their name.*

We can understand why Popper was convinced he had isolated an essential aspect of experimental scientific practice with the theme of falsification. He had seen clearly that the challenge of falsifiability, and thus the possibility of making it a principle, was crucial. What he had certainly seen less clearly was that it was not a question of a decision that a scientist would be free to make with regard to a theoretical proposition. Likewise, with the notion of a "conventionalist stratagem," he had clearly seen that what scientists defined themselves against was the power of fiction. What he saw less clearly was that the possibility of speaking of a stratagem—that is, of denouncing this power—was itself also dependent on the counterpower created by the experimental apparatus. From the viewpoint inaugurated by Galileo and his successors, the power of fiction reigns wherever experimental invention has not taken place, despite good wills and heroic decisions.

If we had to define the new type of "truth," for which the mathematical definition of movement is serving us as a prototype, what we must think of, rather than the famous distinction between how and why, is the idea of a *negative truth*: a truth whose primary meaning is to resist *the test of controversy*, unable to be convinced that it is no more than one fiction among others. The "authority" of experimental science, its claim to objectivity, thus has *no other source than the negative*: a statement has conquered—at a given epoch, of course, and not in the absolute—the means to demonstrate that *it is not* a simple fiction, relative to the intentions and convictions of its author. But it is not differentiated from the fiction by anything other than its power to silence its rivals.

The experimental statement is thus *mute* with regard to its positive scope. It is all the more so in that one's rivals, who are constrained to fall silent, are not just anyone. They are those who accept the situation of controversy, that is, the challenge of the experimental apparatus. Galileo's apparatus, for example, is powerless to silence those who would refuse to consider the movement of moving bodies to be of any interest, for whom understanding movement first of all means understanding the growth of plants or the gallop of a horse. The latter is "excluded" from the laboratory, from the place where rivals gather around an experimental apparatus that could put it to the test. But the process of selection-exclusion is not enough to establish the difference between "scientists" and "nonscientists." There is no other criterion for this than the dynamic of the scientific fields that are constituted in producing this criterion. It is a process that must be *followed*, in the sense that it is both the stake and the product, a creation of the collection of "colleagues" whose objections, criticisms, and interests are recognized as pertinent.[1] Others, such as the philosophers and historians, whether they accept it or not, remain "at the door of the laboratory," and can enter it only in two completely distinct modalities: either by confusing it with a windmill, that is, by denouncing it as arbitrary, which, for the legitimate inhabitants, will only manifest their incompetence; or else by succeeding in having their objections and counterpropositions admitted—a rare event, which will be greeted as a "revolution," or at least an inflection in the course of history.

The invention of an experimental apparatus is what makes Latour's principle of irreduction pertinent: it is an operator applicable to *both* things and humans. It proposes both a staging of things and an operation that disqualifies those, among humans, who do not accept the challenge of this staging. To be comprehended, it must be described according to a perspective that *follows* the perspective of the colleagues it qualifies (a perspective that adopts, by definition, the history and epistemology of the victors), and which thus can always be accused of arbitrariness by the others. This is why any epistemological rationality that requires a norm to justify the history in which the criteria of scientific legitimacy are invented and stabilized can lead straight to relativism, as we have seen in Feyerabend's case: such criteria require, like an anamorphosis, the pinpointing of the perspective (here, that of history) in relation to which they have meaning.

It is all the more important to emphasize that, because experimental statements lack the power to force the protagonists to enter the laboratory, this proposition has an inversely symmetrical consequence. The experimental statement has no positive proof at its disposal that would allow it to establish its own signification and have it accepted *outside the laboratory*, that would allow it, for example, to identify, among the multiplicity of different phenomena that proliferate in the laboratory, those to which it constitutes a path of privileged access. The statement, in effect, has relevance only if the selection of traits brought about by the experimental apparatus is itself recognized as relevant. It *proposes* to judge a phenomenon in terms of an *ideal* (the categories that correspond to the experimental apparatus) and the *distance from the ideal* (the parasitical and secondary effects that complicate the situation, and which one must learn how to manage). But it cannot impose this judgment. Outside the laboratory, there is nothing to prevent those to whom it would like to address itself from claiming that, as regards the field that occupies them, the statement merely designates a fiction, that it is, as Sagredo said, "a definition elaborated and accepted in the abstract." It was in this manner that French "mechanical engineers" protested, during the entire thirteenth century, against the arrogance of the academic "mathematicians" who undertook to subject them to their "laws," in both senses of the term.

In other words, the experimental event does not constitute a response without also posing a problem. It does not create a difference between those it gathers together and those who remain indifferent to it without also posing the political question of knowing if and how this indifference will be broken, if and how the consequences of the event will be propagated outside the laboratory. The experimental event makes a difference, but it does not say for whom this difference has to count.

Concerning those who have accepted to gather around the experimental apparatus to recognize its possible relevence, we must first of all say that they have allowed themselves to become *interested*. Getting people together in a laboratory is not a right. Anyone who thinks he has this right is recognized as a "mad scientist." He moves forward alone, armed with facts that, according to him, should logically bring the general assent to his side. He demands that we take them seriously, as recommended by epistemological treatises, and becomes indignant, in the name of the values of science, when his proposition is not recognized as scientific. But we also know of disciplines that cannot be made to admit that they can produce something other than fictions. This is the case with parapsychology, which, since the founding of Joseph B. Rhine's laboratory in 1930, has devoted all its efforts to inventing ever more rigorous experimental protocols, but it runs into "non"-interlocutors, ready to admit any hypothesis the moment it allows them to conclude that there are no facts. The rules of scientific controversy founder: critics refuse to let themselves become interested, to gather together in the laboratory. They are content to recall some cases, supposedly valid for all, in which "everyone knows" that there was only an artifact, in the negative sense, or trickery.[2]

This example, one among many, shows that the simple opening of an experimental controversy is already a success: a statement has succeeded in interesting colleagues who are equipped to put it to the test. "To let oneself become interested" is the prerequisite necessary to every controversy, to every putting to the test.

This is not at all surprising, for letting oneself become interested is a risk. Interested scientists are scientists who ask themselves if an experimental statement can intervene in their problematic field: What difference will it produce? What new constraints and what new possibilities will it determine? In short, what signification will it be able to take on? Accepting participation in a putting to the test thus not only means accepting the possibility of a new practice, in the sense of a simple new instrumental possibility; it means accepting the possibility of a new *practical engagement*. Experimental procedure, truth, and reality will quite possibly enter into a regime of new mutual engagement. It is indeed "engagement" we are speaking of here, in the *aesthetic, affective, and ethological* sense, for the three articulated terms— procedure, truth, and reality—only come together in the mode of a new way of existing and making exist, *where the procedure produces truth with regard to a reality that it discovers-invents; where reality guarantees the production of truth if the constraints of the procedure are respected; where the scientist submits to a becoming that cannot be reduced to the simple possession of a knowledge* (which Kuhn had clearly seen). This is why interest, as the sensibility to a possible becoming, is what an innovative scientist must try to create—a question of life or death.

Authors to Interest

Authority and author, it will be recalled, have the same root, and it was medieval, so-called scholastic, practices that bestowed on them their solid significations. "Authors," in the medieval sense, are people whose authority comes from a text, which can be commented on but not contradicted. This in no way implies a practice of submissive reading, on the contrary. Thus, in Thomas Aquinas's *Summa*, the authors called upon as witnesses to a determinate question do so in the form of citations abstracted from their context. The game and the stake [*le jeu et l'enjeu*] are to make them agree while, most often, holding to the letter of the citation, without discussing the meaning ascribed to it by the author. In other words, the author creates the "authority," but Thomas makes himself the judge and treats the author-authority as a witness called upon for comparison. Thomas has to assume that the witness spoke the truth, and his judgment will have to take his testimony into account, but it is he who actively decides the *manner* in which this testimony will be taken into account.

Thus, the difference between scholastic practice and scientific practice is not as radical as we might have thought. Thomas recognizes that "authors" create authority, but he acted as if he knew himself to be free to determine the way they must be taken into account. Scientists recognize "nature" as their sole "authority," as the phenomenon they are concerned with, but they know that the possibility for this "authority" to create authority is not a given. It is up to them to constitute nature as an authority.

The great difference, in fact, comes from the link between authority and history. The scholastics try to make various authors agree—pagan philosophers, Christian doctors, and the divine author of revelation. Their ambition is to stabilize and harmonize history. In the sciences, *to succeed in constituting nature as an authority and to make history* are synonymous. The power to "make the difference" is on the side of the event—the creator of meaning, but still waiting for significations. The laboratory where a new experimental apparatus withstands the tests that will make it recognizable as something capable of giving a phenomenon the power of bestowing authority on its representative is mute with regard to the fields in which this representative will have the title to speak. In other words, the event poses the problem of its effects, and gives meaning to the history that, alone, belongs to the response.

In this singular link between authority and history, we can see the principal characteristic of the "politics" invented by the sciences: the flaunted solidarity between what Aristotle distinguished as *praxis*, whose virtue was *phronesis* or practical wisdom, and *poiesis*, whose virtue was *technē* or know-how. The Aristotelian distinction moved between the work of fabrication, having its end in a product, and human action, which is open and unlimited, because it concerns a field defined by the plurality — the rivalry, conflict, and complementarity — of human beings living together.[3] The laboratory, from all appearances, is the place of *poiesis*, for it is the place where a "fact" is fabricated, a fact whose vocation is to make authority, to constitute the unity of the end (the statement that represents it) and the means (the experimental apparatus). But it is also the place of a *praxis*, for this "fact" is not an end; it inaugurates, as the epistemologists would put it, a "research program" — that is, to put it more concretely, it is addressed to other authors with whom it proposes to "live together" in a new mode.

The link between *poiesis* and *praxis*, between "fact" and "history," is obviously not an absolute novelty. We can retroactively dispute Aristotle's distinction. The novelty is that his link now defines a class of authors who exploit it systematically. It is this novelty that escapes the apolitical conceptions of "rationality" invented by the theoretico-experimental sciences. Whether it is a question of Alexandre Koyré putting the physics of Galileo and Newton under the sign of Plato (mathematical intelligibility of the world), or critics of technoscience putting onstage the "merely operative" character of scientific concepts ("science does not think"), the analogy (with the Platonic vision of the world) or the opposition (to the demands of philosophical or symbolic intelligibility) conceals the change of scene that transforms the signification of words. *Material, electron, vacuum* do not receive an "operational" definition, as if it were enough simply to decide to subject them to an operation; rather, they become that on which *we* are now able to operate, and it is this "we" that is decisive, the creation of a collectivity with which matter, electron, or the vacuum will now make history. It is from the *political* definition of this collectivity that epistemological terms such as *objectivity* or *theory* take on meaning.

Scientific practices imply, correlatively, a *phronesis*, a practical wisdom that concerns the plurality of humans and the diversity of their interests, but in a new genre. This is why it becomes possible to create a scientific imperative from the notion of "interest" without for all that running against an "established sentiment," a sentiment that designates the "disinterested consensus" of scientists as the guarantee of their propositions. Interest is here redefined by the link in which the ensemble of *poiesis* and *praxis*, *technē* and *phronesis*, *fact* and *history* are reinvented together.

Interest derives from *inter-esse*: to be situated between. This not only means to stand in the way of [*faire écran*], but first of all to make a link between [*faire lien*]. Those who let themselves become interested in an experimental statement accept the hypothesis of a link that engages them, and this link is defined by a very precise claim, which prescribes a duty and confers a right. Those who accept it must be able to maintain that they have done so only to the degree that the statement did not link them to an author "like the others," and that this link did not signify a relation of dependence to interests, convictions, or ambitions that would be the clandestine ingredients of this author's proposition. It also means that those who accept to engage themselves, who admit into their laboratory the experimental apparatus authorized by the statement, have the right to preserve their position as independent rivals, and do not have to become disciples subject to the unanimity of an idea. They simply recognize that the apparatus has succeeded in allowing the phenomenon to "create authority," to bear witness to the way it *has to* be described.

The possibility of this redefinition in fact separates the question of the sciences from the group of philosophical readings that have disqualified "interest," and have, in one way or another, grounded their judgment in relation to the true, or indeed in a transcendent order (readings that are the heirs of Plato, the first "professional thinker," according to Arendt and Taminiaux). Interest then becomes something that nourishes the power of fiction, that separates humanity from what, in one way or another, should be its vocation. Interest becomes something that must be transcended, something we must go beyond, something of which we must be purified, and against which we must be converted. The singularity of the sciences, as I have tried to characterize them, is less to break with this notion of interest as an obstacle than to transform it into a stake. Interest in itself is not disqualified. The only thing that is condemned is the failure of the person who, while claiming to interest others, does not succeed in making them admit that their own interests can be forgotten. The future opened by the statement must be available to "everyone," it must create a community of "equal and different" heirs to whom the problem of history is posed.

If scientific practice entails certain tests within a field of immanence — tests that philosophical doctrines have often referred to as the heaven of ideals — it does not, for all that, raise the suspicions that traditionally weigh on the notion of interest. Whereas the True, the Good, the Moral Law, or any other authority transcending interest claims to be able to orient humans in a unanimous direction, to assure their agreement, interests themselves do not have this power. A scientist will not demand that his colleague be interested in his proposition for the same reasons as himself, but only that the colleague pass through the conditions under which this proposition interested him. Moreover, he can himself try to provoke the maximum of *heterogeneous* interests, capable of giving his proposition the maximum of significations. It is precisely *because* interest, as opposed to "truth," does not claim the power to create unanimity, but lends itself to proliferation and association with other disparate interests, that it can bring together authors for whom the event poses the problem of history.[4]

Thus the scientist, as an author, does not address himself to readers but to other authors; he does not seek to create a final truth but to make a difference in the work of his "author-readers." And it is in terms of this difference—in terms of the risks and the promises of history constituted by the statement—that the statement is evaluated and put to the test. Which means, of course, that the scientist is not interested in impartial readers, who would have the same "chance" of getting interested in any proposition, no matter where it comes from and what it implies. Analysts of scientific controversies are perfectly correct in emphasizing the very different manner in which the charge of the proof is capable of being distributed: certain propositions enjoy the benefit of plausibility from the start, whereas other, apparently comparable, propositions do not succeed in vanquishing the wall of indifference. But propositions are not themselves humble defendants, whose sole demand is that they be given their due. For the readers they are addressed to, scientific texts are far from being "cold," given the experiments and conclusions to which they lead rationally. It is a risky apparatus that, indissociably and at the same time, puts onstage both "facts" and readers, and assigns roles to each of them—pertinent critic, incontestable authority, ally, unhappy rival—which it wants them to accept, in a history that it wants to make go through the difference it claims to have succeeded in creating.

The distinction between event and history is in fact derived from the experiment of thought. A scientist is never alone in his laboratory, like an isolatable subject. His laboratory, like his texts, like his representations, is populated with references not only to those who could put them in question, but also to all those for whom they could make a difference. How did Pasteur represent a microbe? As Bruno Latour writes, "this new microscopic being is both anti-Liebig (the ferments are alive) and anti-Pouchet (they are not born spontaneously)."[5] But Pasteur had already envisioned other possible significations, and numerous other practices where his microbes could make a difference. We have in fact multiplied the modes of intervention of microbes in our own knowledges and practices, but the scientific identity of these microbes is always the sum total of what certain authors have succeeded in making them affirm against other authors.

Bringing into Existence

"Microbes exist, Pasteur discovered them." We have to construct a signification for this statement that does not impinge on the Leibnizian constraint I have given myself—not to challenge established sentiments. This does not mean, let us remember, that we cannot challenge the sentiments of people whose position depends on currently dominant relations of force. In this case, I will have to try to describe the impassioned activity of scientists in a mode that does not call forth denunciations, but that makes its specific vulnerability to the temptations of power intelligible. This vulnerability, I would like to show, seems to me to be linked to the passion for making history, that is, for rendering "truly true," discovered and not invented, those beings for whom the laboratory creates the trustworthy witness.

From the viewpoint of constructivist epistemology, the notion of discovery is detestable. It implies, in effect, that the thing scientists are referring to in fact preexisted the construction of this reference as such. Not even America was discovered, some say, but invented. And certainly, it is only from the viewpoint of Columbus and his successors that we can speak of a discovery; the Aztecs did not know they had been "discovered." And the "thing" that was discovered had never been a "preexistent" America, but a multiplicity of interlaced and conflictual Americas, not unlike the interests, significations, interpretations, and stakes that were tied to it, and that captured it within an irrevocable history. But established sentiments can rebel against all this, and emphasize how difficult it is to use a syntax that could avoid presupposing the preexistence of something that would be called, not "America," perhaps, but "an inhabited land, which one had to cross the ocean to reach from Europe." If this land did not preexist, what have we captured in our histories? What are all our interests, stakes, and interpretations tied to?

It is possible, I believe, to say that America was discovered, even from a constructivist viewpoint. Discovery here does not designate an identity between "that which" preexisted and "that which" we now designate as having been discovered, America. It designates the fact that, for us Europeans, not only did America create an event, but it did so *without having any need*, after Columbus's voyages, *to designate the laborious artisans* who would have succeeded in inventing a way to force our interest in it. To be sure, from then on the event referred equally to us ourselves. We know, for example, that at the beginning of the fifteenth century, the

Chinese emperor Yung Lo sent a gigantic flotilla to establish diplomatic relations with the African kingdoms, and that, after his death, the enterprise was purely and simply abandoned. For the Chinese, if not for the emperor, an event analogous to the "discovery of America" did not take place. In what mode did the "external world" exist for the Chinese?

Thus, it is not in an absolute sense, but for Europe at the end of the fifteenth century, that Columbus's voyage can be said to have "discovered America." But "America" shows that it "truly existed" before Columbus by the multiplicity of resources it contained for us, that is, by the uncontrollable proliferation of the consequences of its "discovery." Theologians, kings, storytellers, sailors, merchants, defenders of the Indians, adventurers — there are literally enough for everybody. America imposes itself as a "discovery" not through some sort of adequation between the words we have invented to express it and that which preexisted the words, but by the multiplicity that overflows the words, plans, vocations, dreams, and convictions it had the power to make exist — for better and (especially) for worse, from the point of view of its inhabitants.

What other definition can we give to the reality of America, than that of having the power to hold together a *disparate* multiplicity of practices, each and every one of which bears witness, in a different mode, to the existence of what they group together. Human practices, but also "biological practices": whoever doubts the existence of the Sun would have stacked against him or her not only the witness of astronomers and our everyday experience, but also the witness of our retinas, invented to detect light, and the chlorophyll of plants, invented to capture its energy. By contrast, it is perfectly possible to doubt the existence of the "big bang," for what bears witness to it are only certain indices that have meaning only for a very particular and *homogeneous* class of scientific specialists.

The passion of these cosmologists can be be said to "make the big bang exist," that is, it makes them capable of speaking of it as a discovery. To do so, they must seek to multiply the links between the big bang and those scientists who do not belong to their own specialty. As Latour says, they must multiply the "allies" of the big bang, those for whom it makes a difference, those who need it in order to give meaning to their practice. For what matters when it is a question of "making exist" is less the *number* than the *disparate character* of the allies. The number can merely signify the effect of an unstable and fickle fashion. If the allies belong to a homogeneous class, the stability of the reference only holds for a single type of test. America affirms its existence prior to the discovery of Columbus by the multiplicity of tests to which those who define their practice in reference to it have subjected it.

The task of the laboratory scientist is more laborious, for what he discovers at the end of a test tube is not America. Most often, one creates a new phenomenon. Sometimes, one identifies a new way of addressing a well-known phenomenon, already overdetermined with significations and supported by multiple practices. This is why, most often,[6] one has to *work* to make a new scientific being exist, and scientific "discovery" thus has as its condition a very different history than the quasi-instantaneous explosion of the consequences of the discovery of America, a history in which interests have to be *mobilized*, that is, both provoked and aligned in such a way that they create a link between a being they unanimously designate and the disparate multiplicity of sites where this being is henceforth actively implicated.

The paradox of the scientific mode of existence is that the laborious nature of the construction does not contradict the quest for the "truly true."[7] In effect, this construction is put under the sign of a risk: the allies, capable of bearing witness in their practice to the existence of a "scientific being," will not allow themselves to be recruited "in the name of science"; rather, it is the created reference that must open up their practice to new possibilities. This paradox is analogous to the paradox of the "artifact," which we have already emphasized. To be sure, all experimental facts are "artifacts," but because of this they give meaning to the tests whose vocation is to assess the difference between artifacts — tests that disqualify artifacts that are said to be purely relative to the protocol that created them, and accept artifacts that are said to be "purified" or "staged" by this protocol, which could then, without being destroyed, give rise to other modes of purification, and be put to the test by other questions. To be sure, the beings that science brings into existence are "invented," in the sense that their attributes are relative to our histories. But it is precisely for this reason that their existence depends on the multiplication of histories whose common trait is to refer to them, to designate them as the, if not sufficient, at least necessary condition of possibility for those histories.

Mediators

To speak of "hydrids" that refer both to nature and to human activity, invented by the latter to bear witness to the former, Bruno Latour suggests that we avoid the term *intermediaries* — which implies a problematic of purity, fidelity, or distortion in relation to something that is always already present — and instead use the term *mediators*. What comes first, then, is the activity of mediation, which not only creates the possibility of translation but also "that which" is translated, insofar as it is capable of being translated. Mediation refers to the event, insofar as its possible justification by the terms between which it becomes situated comes after the event, but even more so insofar as these terms themselves are then expressed, situated, and make history in a new sense.

In *We Have Never Been Modern*, Robert Boyle's air pump occupies a place similar to the one I have given Galileo's inclined plane: it is both mediating and, as such, at the center of a conflict between Boyle and the philosopher and politologist Thomas Hobbes, who contests the possibility to which it bears witness.[8] Hobbes rejects the possibility of the vacuum for ontological and political reasons[9] of first philosophy and he continues to allege the existence of an invisible ether that must be present, even when Boyle's worker is too out of breath to operate his pump. In other words, he demands a macroscopic response to his "macro"-arguments, a demonstration that would prove that his ontology is not necessary, that the vacuum is politically acceptable. Now what does Boyle do in response? He chooses, on the contrary, to make his experiment more sophisticated, to show the effect on a detector—a mere chicken feather!—of the ether wind postulated by Hobbes in the hope of invalidating his detractor's theory. Ridiculous! Hobbes raises a fundamental problem of political philosophy, and his theories are to be refuted by a feather in a glass chamber inside Boyle's mansion![10]

Scientific mediation differs from "the discovery of America" in the sense that it constitutes a *work* of redistribution and redefinition whose protagonists are actors subject to the principle of "irreduction": whatever mediation affirms, it is necessary that no one be able to ascribe it to the power of fiction. This means, correlatively, that the work is also political, for it is a matter of defining which protagonists could, in a given case, refer the mediation to a fiction. "Around the work of the air pump we witness the formation of a new Boyle, a new Nature, a new theology of miracles, a new scholarly sociability, a new Society that will henceforth include the vacuum, scholars, and the laboratory."[11]

The existence of the vacuum thus has never been "proved" in the sense that this demonstration would have satisfied the adherents of the ideal of intersubjectivity, the agreement of rational subjects capable of understanding each other and arriving at the stable agreement on a problem, situation, or thing. Intersubjectivity makes the possibility and duty of an agreement rest on subjects, on their "communicational reason," as Habermas would say. Intersubjectivity implies an ascent to a universal form that allows us to situate, understand, and calmly discuss differences; it implies a reference to the truth that, even without content, conserves its traditional power to create unity, beyond divergent interests. Now, no one ever responded to Hobbes's arguments, just as no one, today, tries to respond to the Kantian argument concerning the impossibility of taking the universe as an object for science. "Hobbes" and "Kant" were faced with a violent choice: either they enter the laboratory—Hobbes discovers a weak detector for his ether wind, and the Kantians discover a way of counterinterpreting the residual radiation of a black body—or they fall silent; unless they protest, like Heidegger, that "science does not think."

Existence, in the scientific sense of the term, has very little to do with "intersubjectivity," with the ideal fiction of human protagonists looking into each other's eyes and together managing to bring out the values, presuppositions, and priorities that unite them beyond their conflicts, which henceforth become secondary. Scientists rarely look into each other's eyes. Rather, they turn their backs to each other, each in their own laboratory striving to invent the means of creating a fact that will silence their adversary. Their discussions rarely rise toward a more powerful reference than the one that articulates their conflict,[12] but rather plunge toward the apparently insignificant "details," suddenly reinvented as "capable" of making the difference, capable of constituting a new mediator.

There are nonetheless great differences between the two mediators, between Galileo's inclined plane and Boyle's "air pump," differences that will permit us to make them two different tutelary apparatuses of theoretico-experimental practice.

The inclined plane puts on stage a well-known movement, that of falling bodies. It does not bring this movement of bodies into "existence," but designates it in its novel singularity: it is the movement that will now be recognized as capable of "saying" how it must be described, capable of imposing the articulation between three distinct concepts of speed. By contrast, the air pump produces a drop in atmospheric pressure, which makes the void "exist" as a limit point, corresponding to an ideal pump, but it does not say how the vacuum must be described. Moreover, Galileo's inclined plane can make what it describes vary, as the variables of movement, but it is attached to the falling movement of heavy bodies. The air pump constitutes the invention of a scientific instrument, available for other questions. In this sense, it creates a practice that is the ancestor of what we today call physicochemistry, or phenomenological physics. It does not give the reasons for the phenomenon it creates, but it can be integrated into any situation where pressure, which it constitutes as a variable, can intervene. How do boiling temperature, specific heat, reaction speed, the relation between temperature and dilatation, and so on, vary as a *function* of the variation in pressure?

To this difference between the two events of mediation, there corresponds two distinct "styles," which imply two distinct ways of "recounting" the relations between the new protagonists gathered together in the laboratory and those who, at its door, are suggesting justifications and demonstrations. In this way, the history of the inclined plane in Galileo is most often recounted as the triumph of a path that would find its truth in a mechanist philosophy like that of Descartes. In fact, Descartes did not appreciate Galileo's physics at all,[13] and the "quarrel of living forces," which occupied the first half of the eighteenth century, would oppose the successors of Descartes to those of Galileo, including Leibniz. This does not prevent the Galilean event, invented by Galileo himself,[14] from encouraging a philosophical reading of the resulting science, to which the term *rational mechanics* bears witness: the representatives of reason are not only authorized but invited to enter the laboratory to decipher, in the description of mechanical movement, the categories of objective thought. By contrast, the "air-pump" style celebrates the rupture between philosophers and the inhabitants of laboratories, that is, the capacity for *matters of fact*, for the facts created in the laboratory, to impose themselves *despite* rational arguments. Laboratories here close in on themselves, that is, they exclude those who do not accept the "verdict of facts," and organize themselves into a network, that is, they enter into a history in which the uses of the pump (that is, the mediations between the "vacuum" and phenomena) are made to proliferate.

Let us note in passing that the relationships between these two tutelary apparatuses, the inclined plane and the pump, are themselves also the matter of history, which here concerns first of all not the creation of differences between scientists and "nonscientists," but that between scientists themselves. In this manner, the event "atoms exist," which marks physics at the beginning of the century, celebrates the difference between physicians who go "beyond phenomena" and those we could call the "descendants of Boyle," who wrongly stuck to immediately observable *matters of fact*, and refused atoms as speculative. Just as Galileo put his invention under the sign of Plato, and Boyle put his under the sign of the "fact," the theoretical physicists of the twentieth century put the difference they created between theoretical physics and "phenomenological" physics under the sign of the freedom of the mind nourished by faith in the intelligibility of the world.[15] But neither Plato, nor the "verdict of facts," nor the faith of the physicist allows one to comment on the event in terms of influence or philosophical convictions, to create a continuity or the possibility for the historian of ideas to speak in terms of the eternal return of the "same ideas." They were rather "captured," redefined by the operation that mobilized them in the service of a new history.

One final difference distinguishes the inclined plane and the air pump. The inclined plane no longer exists except in pedagogical laboratories, for its testimony is integrated into the equations of physical mathematics, into the very definition of the dynamic object. This is why no one can deal with Galileo's inclined plane without "rebecoming Galileo," without putting himself or herself in the presence of the apparatus that imposes the way of describing movement that it stages. The air pump has been ceaselessly transformed since Boyle's time. As soon as the signification of its testimony was admitted, this transformation could be described as a "perfectioning." To speak of it as a technical progress is to give oneself the right to call it an "air pump" and to admit that the vacuum that it designates exists. It henceforth constitutes a classical inhabitant of all the laboratories admitting the existence of the vacuum — in the sense, in any case, that it defines the vacuum.[16]

The air pump, once it was recognized as a vacuum pump, has become the typical example of what Bruno Latour has called a "black box":[17] an apparatus that establishes, between the data that enter it and the data that come out, a relation whose signification no scientist would want to contest, for to do so he would have to oppose himself to a disparate crowd of satisfied users and to rewrite entire chapters of numerous disciplines. One can use a vacuum pump while remaining perfectly indifferent to both its functioning and its prehistory. Most of those who use it only know how to use it, and are only concerned with its performance. Its very evolution conveys this vocation: an ever-clearer distinction between what concerns its construction, which is now industrialized, and the user, whose competence is limited to some ultrasimple manipulations and the reading of a screen. In other words, the "vacuum-pump" apparatus expresses a relation of force that seems, or at any rate affirms itself, to be practically irreversible. It designates its users, whether they are scientists or nonscientists, as being incapable of putting its testimony in question, incapable of putting in question the "fact" it establishes. Except for a conceivable but rare exception, the controversy will remain downstream, or will situate itself upstream. Anyone who would want to focus the controversy on the apparatus itself would have the multitude of satisfied uses set against him. He would have to "undo," that is, interpret differently, the multitude of facts of which the pump has been an integral part.

Political Questions

The difference between the inclined plane and the air pump signals the limits of the "political" analysis centered until now on a negative truth, a statement that does not itself have the power of defining its relevance "outside the laboratory." More precisely, we are focusing on a "democratic" mode of description: the production of scientific existence here depends on a history in which the allies to be interested are defined as "equals," freely testifying to the difference that has allowed them to create the link they have accepted—an ideal history, if you will, whose relation with the real practice of the sciences poses as many problems as does the history that unites the democratic ideal with the mode of political management of our societies.

Galileo's inclined plane imposes on us the problem of the hierarchy of the sciences in the sense that its testimony, integrated into the syntax of the equations of mathematical physics, has hitherto prevailed against the testimony of movements—and even, since the end of the nineteenth century, of physicochemical transformations—that seem to require another syntax.[18] The difference between "fundamental physics" and "merely phenomenological" physics has not been admitted without conflict. It is inseparable from a history that constructs an inequality between physicists, a distribution of the rights some of them can claim with regard to the object they represent.

As for Boyle's air pump, it imposes on us the problem of the "product" of scientific laboratories. People who open a packet of coffee and hear "pshhht" know that the container was "vacuum-packed"; and, whether they like it or not, that bears witness against Hobbes to the power of Boyle's pump. The product of the laboratory is a rather different work than the one that produces an alliance or the hierarchization of laboratories. It is no longer a question of excluding or selecting protagonists, but of including, making the event exist for the maximum number of interested parties, whether they are *competent or noncompetent.*

In both cases, to be sure, the problem of power is posed, whether it is a question of the power of a discipline on other fields of knowledge or of the power of the redefinition of social, cultural, administrative, or productive practices. Mobilization no longer simply concerns those who will cause the proliferation of mediators, that is, the attributes conferrable to the reality to which they refer; it also concerns those whose activity will be submitted to this reference, and those who will use it according to the modes of engagement in which the imperative "make exist" changes meaning.

This question of power, however, is not parasitical to the practices of the sciences. It is important here not to bring into play too quickly the opposition between "true science" and "ideology," the first responsible for properly scientific invention, and thus for the history of the sciences as "progress," the other conceived as an "impurity," more or less fatal, but in any case separable from progress. The question of power, as I am treating it here,[19] is one of the "effects" of the event. It corresponds to a question that is posed to the actor-authors who are sustained by this event: What authorizes the difference between science and nonscience that authorizes them? How far will this difference be recognized as a source of authority? In what domains will it simply constitute a constraint for a problem it does not define?

To these questions, all of which are indissociably scientific and political, the notion of the paradigm, for example, gives a version that is all too determinist: as if scientists were free to judge, under the relation of resemblance with their practice, every phenomenon that is proposed; as if these phenomena were naturally available to them without anyone offering resistance to their enterprise; as if they did not have to construct the means of making others recognize that their science bears upon the phenomena in question.

Posing this type of question creates a new perspective on the "autonomy" of scientific communities. Autonomy does not constitute an attribute of scientific practice, any more than does objectivity or purity. It is a practice characterized by far too many stakes. It cannot be presupposed that scientists can be "purified" by what makes them an author. On the contrary, contemporary studies on the practices of the sciences make visible the extraordinary processes of "bricolage" and negotiation that preside as much over the choice of the problem (feasibility, as a function of financial sources, existing or possible; available instruments; alliances existing or to be created, etc.) as over the work properly speaking (modifications of a subject of research, of the apparatus, of interpretation...). Those who study scientists in the laboratory encounter "authors" who have at their disposal all those degrees of liberty that literary analysis reconstitutes, and puts them to work "like Monsieur Jourdain," without knowing the scientific names that correspond to their everyday practice. What singularizes the science is the question: Can this quality of the author be "forgotten"? Can the statement be detached from the one who holds it and those who take it up? Will a scientific statement, if it is finally accepted, then be considered to be "objective," no longer speaking of the person who proposed it, but of the phenomenon inasmuch as it remains available for other work. In the same way, the autonomy of the sciences in no way implies that scientists remain indifferent to the interests of the "nonscientific" world, or that they forbid themselves to exploit financial, rhetorical, administrative, or other resources that the latter can offer them, or that they can themselves actualize. What singularizes science is that no one can say: This hypothesis, this way of treating a problem, has been recognized as "scientific" because it went in the direction of economic, industrial, and political interests. The scientist who validates such interests instead of and in place of a "properly scientific" argument, manifesting the autonomy of science, will be denounced. A scientist who succeeds in making these interests converge with those of his discipline, and profits fully from the resources that this convergence procures for him, will be honored.

With a phrase such as "succeeds in making converge," we are entering the domain in which the sciences alone can no longer claim to define the scene where their histories are created, and where the scientist can pose a political problem to the city. It is from this perspective in particular that the question must be posed concerning the usual hierarchy among scientists, which is translated by the possibilities of publication and financing. This question was taken up by Kuhn, who privileges the "successful convergence" where the categories of a discipline are accepted as determinative "outside the laboratory."[20] We will return to this. But we will already emphasize here that this problem does not oppose the politics of science and politics in the usual sense, but rather associates them. Whether it is a question of the hierarchy of the sciences or the manner in which science comes out of the laboratories, we can always ask if scientists, wherever they extend their authority, have been able to, and indeed *must*, encounter those who were most capable of putting in danger the categories in terms of which they propose to treat a phenomenon. It is equally from this point of view — associating the two "types" of politics — that we could analyze certain components of the discourse on the sciences, to which epistemologists have tried, in vain, to give meaning.

They must, for example, be taken for political operations, which aim at assuring a space of expansion, without risking the whole of the methodological discourses thanks to which scientists efface the traces of the event that authorizes them. Galileo had already claimed — in a Platonic discourse that Alexandre Koyré relied on a bit too heavily — that the experimental apparatus exists simply to illustrate the truth of the facts, a rational truth that he, like a good midwife, will lead Sagredo and Simplicio to recognize as soon as they free themselves from the illusions of the senses and the authority inherited from the tradition. For his part, Lavoisier affirms, in his *Method of Chemical Nomenclature* (1787), that chemists must rid themselves of the imagination that carries them beyond the true, toward fiction, and any qualities that would make them an "author," in order to let nature *dictate* the adequate description. In both cases, scientists put themselves forward as representatives of a "scientific" or "rational" approach that would have to be generally valid, with a scope that is in principle indefinite. This is what epistemologists have tried to decode in vain. In both cases, objectivity claims to be defined as the production of a procedure that, in the end, is objective, and, as Feyerabend has shown, this claim allows scientists to weaken those who could put the validity of their categories in danger, by assimilating their objections to an irrational resistance to objectivity.

If methodological discourse heralds a type of victory that seeks to sustain the forgetting of the question of its own limits, the production of theoretical judgments concerning reality realizes the same operation by other means. From Galileo's "nature is written in mathematical terms" to Jacques Monod's "chance *alone* is the source of all novelty, of all creation in the biosphere," certain conceptual statements produced by scientists have metaphysical resonances. In fact, these are the extreme limits of a transformation of enunciation realized, at more reduced scales, by every theory.

Up to now, I have spoken of the statement, and not theory, in order to reserve the latter term for scientific productions that construct a representation of reality as it exists "outside the laboratory." The task of this representation is to explain and justify the event that constitutes the invention of an experimental practice, and thus to make one forget the contingent singularity of what made this practice possible. Thus, in the 1950s and 1960s, when the coded relations between DNA and proteins are recognized, or when the genetic code is deciphered, these are experimental statements that are proliferating. But when one speaks of genetic information and defines the living being by its "program," it is a question of theory.

When speaking, as I have already done, of theoretico-experimental sciences, it is understood that theoretical production is expected and legitimate within the practice of the modern sciences. Nonetheless, this is not the prerogative of every statement: it can happen that an experimental relation which is recognized to be weak becomes an instrument of measure, without for all that receiving a determinate theoretical signification (as in the case of specters of absorption and the specific emission of chemical elements before Bohr), or else it receives its signification from another theory (as in the case of the chemical "data" in quantum chemistry).[21] Moreover, it often happens that statement and theory, in the sense that I am defining them, are not explicitly distinguished. Many would term theory what I am calling statement; others would recognize what I am calling theory as the "hard core" of a research program, in Lakatos's sense. Still others, if they are opposed to one of the propositions that I am calling theoretical, will speak of irrational ideological claims. The interest of the definition I am introducing is to refer the question of theory, not to a question of its epistemological status, but to the sciences as collective practices, and to avoid any epistemological opposition between a "true" theory, a legitimate theory, and an "ideological" theoretical claim.

According to my definition, a theory is recognized by the *claims* of its representatives. These representatives claim that, in this or that remarkable case, the phenomenon put on stage by the experimental apparatus is not content to testify in a faithful manner, but *has testified to its truth*. Bacteria have testified that, as a living being, its truth was to be programmed genetically. The phenomenon, then, is no longer simply a faithful witness, but becomes an *object* in the strong sense — that is, the experimental categories lose their reference to the experimental staging as a practice, and become categories of judgment, valid in principle independent of the laboratory where they can be put to the test.

The production of a theory, in the sense I am defining it here, does not have to be denounced; it constitutes "another way" for scientists to make history. But it also *proposes* other ways to make history with scientists, and first of all to recount their histories, and those that link us to them, by being attentive to certain questions: How is the double power constituted — the power over things, whose modes of testifying can now be anticipated, and the power over colleagues, who can now be judged, and whose questions can be hierarchized? Many problems then emerge, which are related to the type of narration we can give of history, and to the possible variants of this history. We would now have the means to tackle the question posed by Feyerabend and the critiques of technoscience: the question of the virulent power that the sciences seem to have when it is a matter of destroying what they can only comprehend as "nonscience."

III

PART

Propositions

S   E   V   E   N

An Available World?

The Power in Histories

FROM THE start of this book, I have been careful to dissociate scientific histories from histories that are constructed "in the name of science." I have shown, using the example of medicine, how the imperative to produce faithful witnesses, which singularizes the theoretico-experimental sciences, was able to be transformed. From being a vector of risk, this imperative has here become an order-word [mot d'ordre], defining as an obstacle the singularity of the living body that medicine deals with, its capacity to be healed for bad reasons.

I have also emphasized the difference between a "paradigm" and a "vision of the world," centered on the recognition of relations of resemblance. Now, the history of the sciences forces us to note, here as well, the possibility of transforming a paradigm into a "vision of the world," characterized not by the power to invent problems but the power to disqualify them. In this manner, if the genetic program is the truth of the living—a thesis defended by Jacques Monod in *Chance and Necessity*—the essential thing is the resemblance between a bacterium, an elephant, and a man, all programmed genetically. What distinguishes them can certainly be interesting, but will have to be redefined through the notion of the genetic program. Embryology, the science attached to a trait differentiating the elephant from the bacteria (there is no embryo for a bacterium), was a leading science in the first half of the twentieth century. With the triumph of molecular biology, it became a set of empirical results, scarcely reliable, waiting for the moment when embryological processes will be successfully made to bear witness to their essential relation with genetic information.[1]

Finally, I gave my undertaking the ambition to relearn, with regard to the sciences, the laughter of Diderot, who was able to like d'Alembert and to respect him without for all that letting himself be impressed by him. The mocking laughter of Feyerabend cannot in the same way touch Laplace announcing that there will only be a single Newton because there is only a single world to discover, and Galileo or Newton "in the laboratory," inventing a way to interrogate phenomena, and themselves being invented in the creation of this new link. The prophetic tone of the readers of technoscience, denouncing the reduction of nature to information processing, cannot conceive of the passion of a computer programmer, who, in order to invent the way a situation can become "processable" by a computer, must submit to a becoming that transforms him into a mediator, site of coinvention of the situation and language. "Operative reason" does not have the same meaning when Jean Perrin announces, "atoms exist, I can count them," and when Jean-Pierre Changeux writes, "in theory, nothing prevents us from describing human behavior in terms of neuronal activity."[2]

Examining the way the reference to science changes meaning — passing from a risk to a method, from the creation of a singular relation with the thing to the judgment that constitutes the singularity of the thing as an obstacle, from the celebration of a conquest to the affirmation of a right of conquest — implies a recurrent question: How has the "world" (that is, the set of practical relations and significations that unite humans among themselves and with things) been rendered *available* to the strategies led "in the name of science"? How have those whose activity, knowledge, and significations have been redefined or destroyed been able to exploit this change of meaning? Why did they not protest that, far from being recognized as "allies" (who it is a matter of making interested, recognized by their freedom to evaluate propositions according to the new possibilities they offer them), they have been judged and disqualified?

I have introduced the distinction between experimental statement and theory in order to make sense of this problem. An experimental statement can be upsetting, subverting the landscape of knowledges, connecting some regions, disconnecting others, but it defines possibilities available to everyone, constraints everyone must take into account, but which everyone must be able to profit from, if they invent the means. By contrast, a theory requires that the hierarchization of the landscape of knowledges it is proposing be socially ratified. Such a science, which poses essential questions, is a leading science [*science de pointe*]. Any other science can be useful, for the questions it addresses to the object can prepare the terrain for the leading science. This other science becomes an applied science, subordinated to a purer science as a parasite or a secondary complication.[3] This other science, finally, must be denounced as a parasite, or as ideological, or not objective, for the questions it poses and the witnesses it seeks, if they were taken seriously, would put the theoretical science in question, and would imply that certain phenomena belonging to the field of the theory bear witness to another type of truth. From Jacques Monod's point of view, the notion of self-organization, created by embryologists, was only an irrational survival of old romantic doctrines.

Every theory affirms a social power, a power to judge the value of human practices. No theory is imposed without social, economic, or political power being at play, somewhere. But the fact that it is at play is not enough to disqualify the theory. The past we have inherited is saturated with "good questions," forgotten in the names of triumphant theoretical claims, but also with theoretical claims that have engendered fecund histories, against every moral expectation. "Crime" can pay in science, as elsewhere. The distinction between experimental statement and theory does not make us administrators of justice, but it does allow us to interest ourselves in scientific strategies, for the past and for the present. A theory can and must be evaluated according to its scope and the effects it aims at. Which are those it means to gather together in a positive manner, in the name of a conviction? Are they already assembled together by an experimental apparatus (minimal scope) or do they include the inhabitants of scientific territories where this apparatus has, until now, produced no difference at all? Correlatively, what kind of appeal do the scientific claims make to general themes such as progress, objectivity, going beyond appearances — which are in themselves signs of an appeal to a "social" power (the public, including nonimplied colleagues, distributors of funding, etc.) — in order to vanquish the skeptics and the unsubdued? Depending on the scope of a theoretical claim, that is, the disparate character of what it means to unify and hierarchize, one can expect that the narrative will become ever more complicated, that more arguments will intervene, that there will always be an increasingly active construction of alliances, always more coalesced interests. The theoretical unity does not unify the network of proliferating interests, it is added to them in the manner of a "judgment of God" in Deleuze and Guattari's *A Thousand Plateaus*.[4]

Interrogated from this angle, two theories can be perfectly distinct while nonetheless using the same type of formalism. For example, the quantum theory of the atom brings together physicists and chemists, all actively interested a priori in the possibilities of their being represented. By contrast, the quantum theory of measure is addressed in principle to humanity in its entirety. It presupposes that everything that exists (and, for example, the famous "Schrödinger cat") can be represented in the manner of a (isolated) hydrogen atom, and it then poses, in as technical a manner as you like, the question concerning the emergence of the properties of "our world" (for example, the emergence of a cat that would be dead *or* living and nondead *and* living). It seems, then, that the very existence of the world in which we live is subject to the "judgment of God," and depends on the verdict of quantum mechanics, subsuming and unifying the ensemble of knowledges about the world. When it is a matter of interesting the public in quantum mechanics, it is obvious that the popularizers will pass through Schrödinger's cat rather than through the hydrogen atom.

We can laugh at "Schrödinger's cat," and follow with amusement the way something that was, for Schrödinger, the illustration of an insufficiency of quantum theory (it does not give an account of the properties of the observable world, of what a cat *must be*, dead or living) has become a symbol of the power that quantum mechanics would have to place in question the evidence of common sense. But can we laugh when doctors affirm that something that is, for the moment, an obstacle to progress in medicine will one day be conquered? What knowledges and practices will be destroyed, or prevented from being invented, in the name of what must be called a "mobilizing belief"—namely, the faith in a future where the body will show that its rational representatives were indeed right, and will permit them to sweep away the claims of charlatans, just as astronomy has permitted the claims of astrologers to be swept away? The laughter is insufficient, of course, but it is necessary. Without it, the forces of the examples from the past and the play of powers that construct the future could combine with impunity, each referring to the other in order to give this future the air of a destiny.

Mobilization

There are many ways to recount the history of the sciences, and to ground the politics of the future on them. What I am proposing puts the emphasis on the event, the risk, the proliferation of practices. What rational medicine requires, for example, grounds on the past the promise of a *reducibility* of what, for the moment, poses an obstacle to it (like the placebo effect). It constitutes in this sense a mobilizing model, which maintains order in the ranks of researchers, inspires confidence in them with regard to the future they are struggling toward, and arms them against what would otherwise disperse their efforts or lead them to doubt the well-foundedness of their enterprise.

We might say, in the manner of Feyerabend, that the production of a mobilizing model is the business of scientists, like the law of silence is that of the Mafia. But before we can say this, we must be able to use other words to describe what scientists do, and it is equally necessary that scientists themselves (like those who leave the Mafia) have the possibility of using other words in order to *betray* their model, if the case arises. To introduce these other words, this other possibility of narrating the advance of the sciences, I would first of all like to emphasize the strange contrast between the effects of experimental practice and the mobilizing rhetoric that takes hold of these effects.

The effects of invention are always the creation of unsuspected distinctions, the possibility of putting into variation what appears as a "given." What is defined as a faithful witness never simply explains what everyone knows, which is something every well-constructed fiction is capable of; rather, it has the possibility to make the phenomenon bear witness to new and unexpected modes, which confer on its representative the power to differentiate this witness from a fiction. Even in cases where a theoretical claim engenders a fecund history, this history does not "realize" the claim without inventing an unexpected signification for it, which transforms the claim rather than simply obeying it.[5] Thus, when Jean Perrin, in 1912, imposed on skeptics the vision of a world in which macroscopic phenomena can be interpreted in terms of events and of movements of imperceptible atoms, he did not impose on them a world reducible to atoms. He imposed on them the multiplicity of situations where atoms, in being decomposed or ionized, and molecules, in entering into reaction and colliding, in determining the erratic movement of a Brownian particle, bear witness to their existence in a mode that cannot be referred to a fiction, for in each case it allows these actors to be denumerated, it allows one to attribute the same value to the famous "Avogadro's number." When molecular biology became capable of deciphering the "genetic code," it thereby became capable of exploding the apparent unity of the gene—the actor in the transmission of heredity—into a multiplicity of interveners, that is, it also invented for each gene a distinct mode of experimental intervention, which makes the transmission vary. Retroactively, we could obviously say that atoms, molecules, and genetic transmission are the given conditions of our history, but they only "make history" (in the sense of scientific referents) by also becoming conditions for *other* histories, transforming what had to be explained in one "case" in the midst of a variety of cases.

Now, the rhetoric that takes hold of the event celebrates the power of reduction. Physicochemical processes are reducible to the play of denumerable atoms; molecular biology has reduced heredity to the transmission of information coded in the DNA molecules. This rhetoric transforms the signification of the "explanation." It is no longer a matter of "ex"-plaining in the sense of "bringing out" *this* aspect of what one is referring to, but also that, and that again—so many "consequences" which in turn testify to the existence of the referent. It is a question of affirming that this referent has the *general* power of leading diversity back to the same. And in this way, it has largely gone unnoticed that the "explicated" diversity does not usually precede the explanation, that it is less a conquest than the product of a practical invention that comes to be *added* to other practices.

The contrast between the proliferation of the new possibilities that sustain the event and give it its signification and scope, and the reductionist rhetoric that is authorized by it, is neither necessary nor insignificant. It translates a staging that makes the invented-explained diversity the guarantor of the general reducibility of a phenomenal field to be investigated—a mobilizing staging that identifies both the conquering army and the landscape defined as available to its conquest. In other words, the staging is not merely rhetorical, but neither can it be identified as an unavoidable consequence of the politics constitutive of the sciences. It constitutes a particular form of political organization, which one must learn to laugh at in order to learn how to resist it, if the case arises.

Mobilization means the making available of the landscape whose properties are denied or identified from the sole point of view of the obstacle they constitute in relation to the ideal of a homogeneous landscape of whose points would all have to be equally accessible: in the Middle Ages, fields were trampled on; today bridges are constructed across rivers quickly enough for the speed of an advancing army to be unaffected. Mobilization means equally the coherence of the whole, an ideally instantaneous transmission between different parts and the central post, which has at its disposal a global image of the situation. (We know that, in Germany, the unification of the local hours had as its principal vector the minister of armies.) Mobilization, finally, means discipline. It is necessary for the different parts to obey the received orders, to become parts of a real body, the responsibility for their activities falling on the single brain that commands them. Any local initiative, even when crowned with success, is suspect.

How to mobilize and align interests without destroying them, without transforming rival interest into an army marching in step? How to discipline scientists in such a way that their local and selective inventions can be recounted in the mode of a conquering deduction, referring the responsibility of the operation to the power authority in the name of which science is activated? How to preserve in members of the scientific community a sense of initiative and opportunity that belongs rather to the guerrilla, but in such a way that this guerrilla imagines himself belonging to a disciplined army and refers the meaning and the possibility of his local initiatives to the order-words of the commander?

According to Kuhn, one can read in the description of "normal science" the invention of this original form of mobilization, which was created in the course of the nineteenth century with the institution of the sites of modern academic research. The paradigm can be deciphered as the operator of this mobilization: it creates a homogeneity of maximal anticipation; it allows every one of its members of the community to invent the way in which he or she will be able to be effectively understood, but it allows the community to make a rapid judgment of these inventions; it invites one to attribute to the discipline the responsibility of success, and to the "incompetent" researcher that of failures; it is transmitted in a largely implicit mode that rarefies what Judith Schlanger has called "cultural memory": the dense copresence of multiple significations, which prevents a wholehearted adhesion to any one of them, and a sense of what other interests have addressed and are always addressing to whatever one is concerned with, which "introduces the world between us and us."[6]

One might wonder if this form of mobilization is not in decline, in at least certain disciplines. The notion of normal science implies a certain slowness, a relative stability of judgments, which constitutes a norm for several generations of scientists. It also implies the event, which aligns interests but creates a difference. It is bothersome from the point of view of the conquering mobilization, between fields where the measure has a signification and a stake, and those where it is an empirical correlation, available for multiple interpretations. In fact, the speed with which new technical tools are today proposed, making their precedents out-of-date, creates a form of mobilization that now has neither the need nor the time to forge a paradigm. To find the means to acquire the last instrument, in order to remain in the running, that is to say, in order to have access to publications that require the type of data it produces, constitutes in many contemporary laboratories an order-word, sufficient to align interests, but without constituting them as the heirs of the event, without the latter sustaining them, inhabitants of a territory marked out by the convictions and practices that celebrate it.

There is a great difference between the paradigmatic mobilization and the mobilization that takes place solely through the speed of technical inno-

vation. The first has the time—in the double sense of the opportunity that constitutes the event and the temporality proper to the invention of its effects—to construct
a representation that could be called "territorial," for it allows one to make the difference between the inner and the outer, to recount the history of the foundation [*fon-
dation*] and the constitutions of grounds [*fondements*], to construct the double dynamic
of "pure" knowledge, authorized by the paradigm, and its applications, bearing witness to its fecundity. The second is lived by many scientists in the mode of dissatis-
faction, nostalgia, and a new sense of vulnerability: data and highly sophisticated
correlations accumulate, but no one truly has the time to think them; the difference
between "before" and "after" is made ever more rapidly, but it no longer applies to
creations that would affirm the autonomy of the territory, but rather to the accelerated obsolescence of the instruments that date the research; the quality of the researchers counts less than their access to the resources that allow them to respond
to the imperatives of the moment; their identity no longer refers to the event that
authorizes their conviction, but to the power of the instruments, which often have
come from other disciplines; it is therefore increasingly difficult for them to resist
injunctions and pressures, ever more insistent, that aim to make them furnish "useful"
information, even if, from their point of view, they have no interest in it. In short,
the menace felt is that scientific research in fact comes to resemble the image given
to it by the "technoscientific" reading—and that, correlatively, the differentiation
between "pure science," focused on researchers' territorial interests, and "applied science," in which these interests are composed with other interests, to the profit of a
double indifferentiation: phenomena that are no longer able to authenticate interests,
because they are made available by disposition by the power of the instrument; scientists who no longer have any reason to resist authorities who would suggest to
them to be interested in this phenomenon rather than another.

      The form of mobilization that describes the functioning of a "nor-
mal science" has been a scientific invention, and it occurred in a context in which
the autonomy of research should have been defined and negotiated no longer in relation to traditional, hostile, and indifferent powers, but in relation to modern powers,
such as states and industries, which are potentially or actually interested in scientific
knowledges and practices. The power of the mobilizing paradigm is equally a *counterpower*, opposed to the threat that research will be subject to "utilitarian" interests.[7]
One can understand the anxiety scientists have when confronted with the precariousness of this counterpower, but one can understand it without, for all that, sharing
their nostalgia. For the construction of territorial disciplines normalized by a paradigm is inseparable from the image of a reductive conquest, which in principle affirms

the availability of what is being investigated. The great mobilizing narratives have always defined progress in the mode of asymmetry: the power of the person who advances in the name of science, and who is distrustful of the "opinions" of those who occupy the territory to be subjected. They have always hidden the fact that, most of the time, not only were the investigated zones not virgin, but the local knowledges, far from being rendered obsolete, were permitted to guide the creation of new pertinencies, which were retroactively described as deductions authorized by the paradigm.

To use a linguistic image, the paradigm affirms the unanimity of the phenomena that speak the same language, but this language is clandestinely enriched with local constraints, which do not figure in the official dictionary, and which must be learned on the spot. To use a geographical image, the paradigm affirms the homogeneity of the landscape, but it conceals the existence of passes and crevasses on the paths that connect the different regions, and it conceals, in the narrative of the official voyage, the aid received from locals, without which the person who arrives could not have invented-improvised a means of passage.[8] The price of this politics of submission of the local to the global is not only a hierarchization of knowledges, systematically privileging the theoretico-experimental enterprise, which alone can arm its practitioners with judgments that mobilize both phenomena and humans; it also ensures a mode of engagement for truth that, situating the truth on the side of power, makes it vulnerable to all powers.

The Patron's Job

Between the constitution of a disciplinary territory and the social construction of a world, which allows the products of the discipline to "make history" with social, economic, political, and industrial interests, there is a relation that is at once intense and masked. This is because a very delicate double movement must take place: the work of disciplinary constitution must exclude and select, whereas the construction of a world that desires, welcomes, anticipates, and gathers must include — or make exist — what the laboratory creates for a maximum of interests, competent or noncompetent.

In three dazzling pages, Bruno Latour allows us to pose the problem of the mode of work and strategy — and not of destiny — in the unenviable mobilization of the world through the products of science. In them, he describes, in the mode of fiction (but without inventing anything), a week in the life of a "patron," the director of a laboratory that has just identified a hormone secreted by the brain, called pandorine.[9]

What is pandorine? It is not an artifact. This we know, for the week described is situated after the controversy that set the patron in opposition to his competent colleagues, who are endowed with a laboratory that allows them to put their molecule to the test. Pandorine—isolated, purified, identified—is indeed a molecule produced by the brain, not a product of contamination or of the degradation of the authentic molecule. Nonetheless, it can be the product of simple honorable research in neuroendocrinology, or the starting point of a "revolution" in the sciences of the brain that wins the patron a Nobel prize; it can remain one biological molecule among others, or be capable of mobilizing, federating, and representing the set of hormones that testify to the existence of a "humid brain" where the "dry brain" of neuronal circuits currently dominates. In short, we do not know what pandorine "is" or how the history of its "discovery" will be narrated, and it is to this problem that the patron consecrates his activity, spending his week traveling, negotiating, speaking, promising, entering into intrigues.

There is, notably, one very promising colleague, who has perfected an apparatus that allows traces of pandorine in the brains of rats to be visualized. The apparatus is a prototype, and the researcher needs the support of the patron in order to interest industry, but if industry were interested, the apparatus could quickly become a "black box," all the more indispensable to the laboratories insofar as the *referees* of specialized journals could demand that all neurochemical research worthy of the name pose the problem of the amount of pandorine secreted for every regime of cerebral functioning studied, and thus render possible the multiplication of its attributes. Then there is the question of editorial committees: the journal *Endocrinology* has not yet recognized the new specialty; "good" articles are rejected by referees who know nothing about it. The National Academy of Sciences would also have to recognize a subsection, without which the members of the new discipline would remain dispersed between physiology and neurology. And at the university itself, a new curriculum would have to attract brilliant young people toward this discipline in full blossom.

The patron is of French origin, and should not France, anxious to share the prestige of this expatriated son, to whom the Sorbonne has just bestowed a doctorate *honoris causae*, make a gesture and relax the rules of scientific politics in order to favor the creation of a truly French laboratory, specialized in researching the peptides of the brain? Already, in the United States, the president is subject to pressure from representatives of diabetics, who have placed their hopes in the breakthrough announced by the patron: they have made themselves his allies in order to demand that he should be given priority, and that the "obstacle" of "red tape" implied by the inevitable clinical tests be alleviated. Other tests are already being discussed with regard to schizophrenics. And, to be sure, the patron has entered into discussion with the executives of a pharmaceutical company: Will pandorine — patented, produced industrially, submitted to clinical tests — be a medicine?

In the middle of his wanderings, the patron announces to journalists that a revolution in brain research is at hand, of which pandorine is the harbinger. But he also exhorts them not to present a sensationalized image of science. And, in the airplane, at the request of a Jesuit friend, he writes an article that links pandorine to the ecstasies of Saint John of the Cross. In a footnote, he announces the death of psychoanalysis.

The patron does what he has to do, if he wants to give pandorine the greatest possible scope, to make it exist in as many registers as possible. This does not mean that this existence depends on his strategies alone. In the laboratories of academic and industrial research, pandorine will have to confront severe tests. But nothing confers on the molecule "in itself," independent of the "patron," the power to provoke these tests, on which it depends in order to impose an interest on other researchers, industries, and scientific journals, without which it would remain a simple molecule, naked, with indeterminate roles and possibilities. By contrast, its demultiplied existence does not limit itself to "dressing up" the molecule in roles and uses, but modifies the landscape of relations that articulate the brain, the anxieties of citizens, the activity of industries, the prestige of disciplines, and the means that are allocated to researchers.

Should the patron be denounced? As Latour remarks, the humble disinterested collaborator, who does not leave the laboratory, is the beneficiary of this apparently interested work:

> she is able to be deeply involved in her bench work *because* the patron is constantly outside bringing in new resources and supports. The more she wants to do "just science," the costlier and the longer are her experiments, the more the boss has to wheel around the world explaining to everyone that the most important thing on earth is her work.[10]

The patron is constrained to be interested in the world, to transform it so that this world will make his molecule exist. He does what he has to do if he wants pandorine to exist, and he does it with great talent. Our researchers are not always naive choirboys, and those of whom we retain the name have most often, and rightly so, proved themselves with their fearsome strategic capacities. But these capacities themselves refer to stratifications of this world, where very different interlocutors coexist with each other. With some, negotiations are "hard" — industrial labora- tories, notably, will not let themselves be taken in. With others — the journal *Endo- crinology*, the Academy, and the university — it will be a matter of organizing a "lobby- ing" activity. Still others, the representatives of diabetics, are used as levers: the suffering of the ill is a fearsome argument, and when patients themselves are re- cruited in the name of hope, decisions can go up "to the highest level," short-cir- cuiting the usual networks where research priorities are negotiated. Journalists must be kept in their place: they must spread the news of the future revolution, without forgetting that the patron is a disinterested scientist, who warns them against any sensationalism. Finally, those who, in one way or another, are interested in human subjectivity must know that the progress of science will sweep away any false differences between "laboratory science" and the "human sciences." Psycho- analysis is ritually put to death, and Saint John of the Cross announces that it is not only intelligence that should be investigated, but also the emotional life. The claims of the patron, here, will entail no putting to the test. His aim is not to gather to- gether his colleagues around a mystic in ecstasy, who has become the faithful wit- ness of the pandorine acting within him, but to disturb, to appear, like Jean-Pierre Changeux and so many others, in the role of the menacing and scandalous represen- tative of the laboratory, whose reductionist advance is authenticated by the protests of the representatives of knowledges doomed to disappear.

The singularity of the patron refers less to an identity of science than to the freedom with which he can construct the triple territory in the name of which he transforms the world: the molecule, the future science of the "humid brain," and experimental progress dissipating the irrational darkness. Nothing seems able to stop him, or to make him see, for example, that at such and such a point, "science" stops and "propaganda" begins. One respects him or one fears him. Journalists, if they snigger, can do nothing about it. The Jesuit journal welcomes with gravity this "summit" encounter between the height of the rational and the height of the spiri- tual. The ill are ready to make common cause with anyone who gives them hope. Psychoanalysts, no doubt, will protest that, far from being dead, they represent "this human suffering that positive knowledges can never hear but can only silence." Even the scientific colleagues of the patron know that a disciplinary reorganization is at hand, which will impose on them new constraints and new demands. It will no doubt be necessary, even if one is skeptical, to find the funds to acquire the new pandorine detector, and to produce figures for its subject that will possibly be without interest. It will also be necessary to have articles accepted in the new subsection of the journal *Endocrinology*. Some of his colleagues will complain, in petto, of the drift of their science toward a simple instrumental practice, but where will these possible doubts get a hearing? How to resist someone who announces a brain available to progress, without inspiring questions that are dangerous to the public, to the ill, and to financial donors?

       The patron does his job as a scientist, he makes proliferate the potential identities of pandorine, and the possibilities of history that will make it exist, if the case arises. And the sole index that he is not ceaselessly changing milieu, passing from a biochemical pandorine to a cultural pandorine, from a pandorine federating a new discipline to a future miracle medicine pandorine, from a media pandorine to a pandorine attracting students destined to do the leading research, is the qualitative difference between the arguments: from close-fought negotiation to rhetoric. As if, this time, we are indeed dealing with a radical asymmetry. The patron recruits allies for his laboratory, which itself represents the neurochemistry of the brain, which in turn represents scientific progress, but some of his allies are defined by demands that must be satisfied, others by a competitive logic to which they must indeed be subjected, and still others by beliefs, fears, and hopes that must be maintained. Correlatively, the different attributes of pandorine are constructed in accordance with the different constraints. Those who link it to hard-to-please allies will eventually be conquered at the price of continual remodelings, which will make pandorine exist in a mode the patron knows he is unable to foresee. By contrast, the pandorine that comes from the laboratory, "naked" but already interesting, thanks to the patron, is in itself sufficient to begin operations of disciplinary reorganization, and to function as a reductionistic war machine, claiming to gather together in itself a multiplicity of available traits, since it pertains to knowledges or practices that laboratory science defines as dedicated, in principle, to reduction.

       Moreover, the hard-to-please allies of the patron have every interest in participating in this asymmetrical construction. The economic profitability of the future detector depends on it, as does the fame of the new generation of medicine that will, one day, perhaps appear on the market. Like the patron, the primary preoccupation of these allies is to "make exist," but existence, here, depends on other tests, which integrate legal, commercial, and economic constraints. They also imply an authority that, officially, does not intervene in the scientific controversies, namely, the public, who need to be turned into consumers. But this is a difference one has an interest in smoothing over. It would be better to respect and maintain the thesis that industry is here a simple intermediary actualizing the secondary benefits of fundamental research, because it is in the name of this thesis that the patron captures the public's interest, impresses the doctors who prescribe drugs, induces the demand of patients—in short, creates the market.

Pandorine is a fiction, and any resemblance to the way true scientists (for instance, those who work at decoding the human genome) come out of their laboratories is purely coincidental.

The Politics of Networks

How can we avoid referring the landscape of our practices, our actions, and our passions to a global authority that would have the power to explain it, and which it would be sufficient to denounce? Bruno Latour is not content to refuse the terms of rationality, efficacity, calculability, and scientificity, all of which explain the construction by the attribute that has succeeded in making us recognize what has been constructed. He also refuses to explain this success in terms of "power." And he is right, if the reference to power has as its vocation to make us forget the network of local alliances, those, for example, that the "patron" tries to create in the name of pandorine, to forget the crowd of mediators, their representatives, and the test they submit to, in order to organize the whole under the sign of a coherent and all-powerful mega-project. Power, when it grows a capital letter, transforms the rhizome into a tree: each branch is "explained" by its relation to another branch, one closer to the trunk, and indeed to the roots, that is, to the site—occupied by a "logic" if not by actors—from which all the rest can be denounced as puppets, acted on beyond their intentions and their plans.[11] The patron, to be sure, does not know what he is setting in motion, any more than do the researchers, who, in order to nourish their researches, nourish in the public the hope in a future where "genetic illnesses" will be curable. But he does everything he can, within the degrees of freedom available to him, and there is no "beyond" from which what, for him, is an initiative could become deducible.

However, it is difficult to put, as *We Have Never Been Modern* sometimes seems to encourage, the "error of epistemologists," rather than power, in the role of the thing responsible for everything that does not go well. Certainly, epistemologists, philosophers, and other thinkers of politics and the social field distinguish themselves by their distrust of hybrids, by their assimilation of mediators to intermediaries, designating the society and/or nature as that which explains them. But "error" does not have to be any more denounced than power. It explains nothing, except insofar as it is a product of the network, characteristic of the *style* of the network that belongs to our epoch, and of the political problem it poses.

Is it the epistemologists' fault if most scientists speak several languages, those they reserve for their colleagues and financial donors, and the one they use when they address the "public," who are defined as incompetent? Is it the philosophers' fault if they learned at school that science would decipher the "laws" that characterize phenomena "objectively," and that their task would be to try to think this situation? Is it the sociologists' or political theorists' fault if sociotechnical innovations or the decisions they comment on are always presented under the sign of an inseparability between what is (the constraints one must take into account rationally) and what must be (the choice subsisting among these possible preconstraints)? To be sure, one can reproach them for a certain laziness, a certain conformism, a misplaced respect. But one must think the network insofar as it arouses, in certain places, the heroic necessity to be neither lazy nor conformist nor respectful, so as not to be taken in.

Error does not emerge just anywhere. In fact, it emerges at points where the negotiations stop, where speech is no longer addressed to actors, who refuse to be taken in any longer, but to those who discover that, because of that very fact, they are now defined as "incompetent"—those of whom one speaks, those of whom one speculates concerning their beliefs, desires, fears, and demands, but in the sense in which they are defined as "influenceable," strategic materials and not protagonists of a strategy. Those who make an error simply commit the error of trusting the rhetoric addressed to the public, to students in the schools, to readers of popularizing magazines, and of not realizing that, like the latter, they have access to a kind of "information" that reduces them to impotence.[12]

To be sure, it regularly happens that one is "mistaken." For example, those who insist on emphasizing that consumers are not powerless, subject to the power of the supply, should recount the numerous stories of products that were refused or whose meaning was altered by consumers, of commercial strategies that were redefined, of unforeseen demands that were urgently satisfied. The political question—that of the difference between qualified actors and the others—does not imply the omnipotence of the first, or the subjected passivity of the second. It is marked in the words that express this type of situation: the public is unpredictable, its reactions always surprise us. These words belong to the same kind of register that would comment on meteorological phenomena. They effect the distinction between those who, actively, seek to foresee, to determine pertinent variables, to articulate them in accordance with the constraints that render decidable what will remain a fiction and what will experiment with possibilities of existing, and those who, through their reactions, will refute or confirm the calculations of which they are the object.

Power is not "beyond" the network, like a truth that would save one from having to follow the construction of ramifications, and would allow one to deduce it. But it qualifies the network and sets limits to it, that is, points where the notion of interest changes meaning, where one ceases to address oneself to the protagonists one is trying to interest, and where strategies presupposing that an interest can "command itself"—or at least be treated as such—begin to appear, which opens the strategies to risks and perils. Such points are numerous, and they draw tangled borders, which must themselves be mapped out. They do not separate things in two, but rather create differentiations [*dénivellations*]. They appear whenever there emerges, as a referent, a relation between two positions: an authority to which is attributed the power to determine its own effects (unless there are obstacles), and a world that is potentially available for the deployment of these effects (unless there are resistances).

The hierarchy of the landscape of scientific knowledges, the role of the model of the theoretico-experimental enterprise, as well as the strategies for mobilization—which never cease to select what the "good" approach is, what the "not yet vanquished" secondary difficulty is—indicate that the differentiations of power traverse the scientific field. But they do not pertain to science alone. The de-levelings themselves form rhizomes. How much easier it is to utilize a scientist already habituated to thinking that his approach "commands interest"! How much easier to handle are scientific experts delegated by a field that is governed by a mistrust of anything that cannot be reproduced in a laboratory! How much more apt to transmit scientific invention as "making authority" are those who have learned it in the mode of evidence! Finally, who would be more prepared to justify, in the name of science, the passage to existence of a sociotechnological innovation than those whose impassioned activity is precisely to "make proliferate" or "bring into existence," for the maximum number of protagonists, the difference between fiction and the faithful witness it has created?

The sciences are not, by destiny, the allies of power, but they are, by definition, vulnerable to all those who can contribute to the creation of differences, the stabilization of interests, the disqualification of annoying questions, the facilitation of the product of laboratories. The singularity I have suggested that we attribute to the sciences—inventing ways to vanquish the power of fiction, and submitting the reasons that we invent to a third party capable of making the difference between them—renders them technically bound up with an "engagement for the true" that defines what is not scientific as merely fictive, available to the putting to the test. This singularity poses the political problem of its coexistence with other actors, for whom the terms of submission and availability have a completely different meaning, who do not address themselves to rival and interested authors but to a world conceived of as a field of maneuvering.

Why is the denunciation of an "operative rationality" so convincing—a rationality that would characterize science and would have systematically destructive effects once it comes out of the laboratories to attack the world? Why are we, and scientists themselves, so often inclined to oppose the scientific, or rational, position of a problem to its "subjective," "cultural," or "psychological" aspects, which must be taken into account, apparently, in another register? Is it not because the same mobilizing strategy prevails "outside the laboratory," in the landscape of human practices, as it does in the strategy of knowledges—the disqualification of anything judged to be an "obstacle," and the systematic privilege accorded to anything that allows one to affirm the power of a procedure?

We must here remember, as an emblem, the end of the thirteenth century when Étienne Tempier proclaimed, in the name of divine omnipotence, the invincible power of fiction. Who was speaking through his mouth? A church careful to re-create the instruments of its authority faced with the rival authority of pagan knowledges, no doubt. But how should we understand these instruments themselves? Just as philosophy, according to Deleuze and Guattari, was not a friend of the Greek city where it was born, and just as science is not the friend of capitalism, the church of Tempier was not the friend of the merchants who, at the time, had learned to define the world, no longer in reference to an intelligible order, but in reference to the possible: a transformable world, a field of maneuvering and speculation. If, as I have tried to show, the reference to "modern science" was born from the invention of the means to get around Tempier's interdiction, it did so not from the perspective of a "return backward" toward a world capable of imposing its reasons, but through the discovery that the power of fiction, the invention of the laboratory, can itself be turned against the arbitrariness of fiction. But the bypassed interdiction can find itself thereby reinforced: it can be in the interest of science to refer to the arbitrariness of fiction *everything that is not science*. It is necessary, therefore, to think of the definition of a "world available to fiction"—which seems to bring together merchant practices, then capitalist practices and scientific practices—in terms of connivance. Between the two types of practices, there is not a hidden identity, which would transform their complicity into a destiny, but a convergence relative to interests, posing a political problem that can receive very different solutions.

A priori, nothing prevents us from conceiving of scientists as being conscious that, in changing milieu, no longer addressing themselves to colleagues but participating in the invention of innovations that are irreducibly social and technological, they must equally change their "ethical-aesthetic-ethological" style. For everything changes when one leaves the laboratory, the place where phenomena are invented as faithful witnesses, capable of making the difference between truth and fiction. Those who were gathered together in Galileo's laboratory, for example, were those who accepted to be interested in the movement that the inclined plane invents and stages. Outside the laboratory, one finds friction, wind, the irregularity of soils, and the density of milieus—everything whose elimination allowed Galileo to establish authority. And one also finds a world acted on by other actors, pursuing other projects, which also imply a differentiation between what must be taken into account and what it would be suitable to neglect. With regard to these actors, the scientist conscious of changing milieus could ask himself: "Why am I so interesting to them? Where are all the others, who are capable of taking into account everything that my laboratory must eliminate in order to authorize me to speak?

No one will propose, usually, to ratify the elimination of the wind if, for example, it is a question of constructing a bridge. Here, the laboratory ideal must come to terms with the "force of things," for the consequences of negligence will be paid for in a mode that clearly makes the difference between success and error. Similarly, every industry is constrained to take into consideration a set of recognized risks, evolving with legislations and regulations, that is, to make the legitimate representatives intervene in the aspect of the problem that the risk designates.[13] But scientists who know that, in coming out of the laboratory, they change milieu and have to change practice, would not expect the law to constrain them not to be unaware of what their laboratories eliminate. They know that the style which belongs to the risks of the test—the invention of ways of purifying a situation in order to make it constitute a faithful witness—changes meaning when it is a question of choice bearing on irremediably concrete situations, where words, if one is not careful, have the power to disqualify, silence, or ratify the amalgams and confusions, that is, to function as slogans.

Such scientists would define as "rational" the necessity that, with regard to a problem "outside the laboratory," all those who are open to representing and exploiting the dimensions of this problem, which they themselves are not taking into account, should be systematically sought out and gathered together. They would judge that it is part of their scientific, ethical, and political responsibility to affirm the selective character of their knowledge, and to demand the gathering together of all those who can contribute to the invention of a pertinent way to pose the problem. They would also know that, once this is done, they must struggle

against the fictions of power, against judgments that disqualify certain interests, turning them into obscurantist obstacles or unacceptable demands.[14] And above all, they
would know that, when it is a question of social becoming, the difference between
success and failure *does not have the power* to decide the pertinence in one's choice of
experts: contrary to the bridge that collapses because of erroneous calculations, a
"social" solution is rarely demented by its effects. Simply, among these effects, it is
often necessary to count on the fact that what has not been taken into account will
become becoming, despairing, clandestine, or ravaged ... and by this very becoming
will confirm the disqualification of which they are the object.

      The difference between these scientists and those who, today,
allow themselves to be selected as the legitimate representatives of a problem, without asking themselves where all the others are and what means have been accorded
to them to assert their competence, does not point to some sort of identity for science,
but to the scientific identity constructed by a mobilized science. Mobilized scientists
will be happy and proud to see themselves called on as experts by a power that recognizes them as the sole legitimate representatives of a problem. They have learned to
mistrust, as an obstacle that has "not yet" been reduced, anything that their laboratory cannot yet take into account, and they will also find it normal that whoever
gives them the means to leave the laboratory also defines, if the case arises, these
dimensions of the problem as negligible, irrational, or destined to sort themselves
out. For them, the essential thing is that the value of their research be consecrated,
and (in the end) receive the financing it deserves. And they will actively discourage
colleagues who have "states of the soul," who try to imagine the "possible" consequences — not represented "scientifically" — of what they are working on. Jean
Bernard, president of the French committee on ethics, "reassures" the public when
Jacques Testart dares to emphasize the dangerous and uncontrollable consequences
of the techniques of artificial procreation.[15] Daniel Cohen, director of the Généthon
program, today disqualifies as "irrational" the concerns of the same Jacques Testart
regarding the social, political, and subjective consequences of certain methods of
genetic diagnosis, and, to the questions posed by researchers in the human sciences,
he opposes the distinction between those who are devoted to battling disease and
relieving suffering, and those who complicate their efforts through obscurantist fears.

EIGHT

Subject and Object

What Singularity for the Sciences?

THE INSTRUMENTS of analysis I have given myself up to now are insufficient, and this insufficiency is made clear by a consequence that, from the political point of view, is very disturbing. I have focused my description on theoretico-experimental practices, as if the definition of the singularity of science—inventing means to make the difference between fictions—were confused with the production of the faithful witnesses created by laboratories. The disturbing consequence is the apparent impossibility of addressing oneself to scientists other than from the point of view of their vulnerability in relation to power. They would have to impose limits on their passion to "bring into existence," and recognize their responsibilities with regard to their choice of allies, who give them the means to fulfill this passion.

It is never good to define a group by a contradiction between its immediate interests and the ethical and political interests to which it must be subjected. The scene is too dramatic, and does not prepare one for laughter. By contrast, it is interesting to transform an apparent contradiction into a tension that already inhabits the group in question, provoking divergent interests in its midst. Certain aspects of the ethical or political demand are then capable of becoming internal stakes, vectors of invention and not motifs of self-limitation.

Other disturbing consequences still follow from the quasi-identification between science and theoretico-experimental science, which I have in fact accepted up to now. One could in effect be tempted to use it in order to settle, once and for all, the question of the scope and authority of the sciences. It seems that science exists only when it is able to invent an apparatus that is able to silence rivals, to institute a situation of putting to the test, where the stake is the power to represent. This possible definition of science is all the more acceptable to many practitioners of the theoretico-experimental sciences in that it hardens the opposition between "science" and "simple opinion" that is presupposed by the experimental staging. Outside the verdict of the apparatus, there are no differences whatsoever, only the noise of indefinitely variable and arbitrary opinions. This definition is thus reduced to impotence once it is a matter of discussing sciences produced outside the laboratory. For example, it has effectively favored the thesis of the American "creationists," who refuse to see the Darwinian narrative substituted for the biblical narrative of the creation of the species. The creationists have complained that the science of evolution cannot claim the title of science, because it cannot boast any of the traits that are manifested in the invention of the theoretico-experimental power. And, moreover, when it comes to pseudo-experimental sciences systematically producing artifacts, this definition of science provides no other means than derision and denunciation.

If the historical problem posed by a contingent process is that of its contingent recommencement, with other givens, it is not contradictory to affirm the primordial character of the experimental event while contesting the hierarchy of the sciences grounded on the theoretico-experimental model. It would then be a question of trying to "extend" the singularity of scientific practices, which were invented by the experimental sciences, to other fields, that is, to also separate [*délier*] this singularity from the invention of a power, from the invention of the means to create faithful witnesses.

The invention of a rather abstract singularity, in order to be separated from the terrain of its birth, must not be confused with the search for a "new science," or, for example, with the search for a "holistic" science, respectful of the world as it presents itself, seeking to reconcile and repair cleavages and conflicts, which we today hear over and over again.[1] In the perspective I have proposed, scientific activity integrates a form of polemic and rivalry, it promises an "engagement" that joins together interest, truth, and history in a mode which is neither that of traditional knowledges nor that traditionally associated with the feminine image: full of softness, conciliation, respect for the feelings of others, faith in a fragile but profound intuition. This is why I have emphasized the interest of Sandra Harding's proposition associating the struggle of the feminist movement with the contrast between the impassioned activity of Newton and Galileo, on the one hand, and the discourse on method and objectivity that authorizes them, on the other. If the "anti-polemical" image of woman had to be veridical, it would have as a consequence the self-exclusion of "true women," those who would correspond to the image, from the ensemble of heirs of the event of "the creation of the modern sciences," which would then be associated with a "male" conception of truth. But my position engages me in turn. I will have to show that the singularity I am proposing for the "modern sciences" effectively separates truth and power, and does not ratify the thesis of the "great division," in the name of which we recognize that, unfortunately, traditional knowledges are condemned by the mere existence of modern knowledges, earthen pots against iron pots.

The challenge I have given myself, to separate science from power without for all that separating science from polemics, can be repeated in the language that distinguishes subject from object. The classical conception of subject and object is the product of a polemical division. The "free" subject is the subject that has purified itself of opinion, once and for all. It knows it is not concerned with objects, whose mode of existence is absolutely distinct from its own. It knows how to relate to these objects, at least in the sense that this relation must have nothing in common with the way it relates to another subject. In one way or another, power, initiative, and questioning are on the side of the subject, with the object being on the side of the "cause," that which subjects discuss and pass judgment on.[2]

The classical distinction between subject and object presupposes power, of course—the power of the subject capable of summoning the object to the tribunal where its cause will be discussed. The laboratory, where the conditions of evidence for the object are defined and where the object is put to the test, is the figure par excellence of this tribunal, the place where the defendant is understood according to categories that will allow one to pass judgment. We can even go further, and say that the "experimental tribunal" is the site where the classical distinction between subject and object was *stabilized*, whereas philosophical discourse, notably that of Kant, attributed to it a general scope.

From the perspective where experimentation is affirmed as a singular practice, which does not presuppose but *creates* both subject and object as well as their relations, no version of these relations, no matter how purified, can claim a general validity any longer. Correlatively, the question of what can become of the distinction between subject and object in scientific practices that are not centered on experimentation is no longer a philosophical question but a question immanent to the sciences, that is to say, a practical question.

In order to separate science from power, is it necessary to contest the distinction between subject and object, or it is necessary to modify it? The thesis I will defend in this chapter is that the singularity of the modern sciences implies the maintenance of the distinction, for it is from this distinction that the risk is born.[3] Once it is a question of science, all human statements *must* cease to be equivalent, and the putting to the test that *must* create a difference between them implies the creation of a reference they designate, which *must* be capable of making the distinction between science and fiction. Thus, the distinction between subject and object, insofar as it expresses this relation of putting to the test, cannot be purely and simply eliminated.[4] The question of knowing who must submit to the putting to the test, however, remains an open one. This question joins up with Sandra Harding's thesis concerning the link between "objectivity" and the critical putting in question by scientific practices themselves, and the relation between the "social experience" of scientists and the "types of cognitive structures" their procedure privileges. It preserves the distinction between subject and object, but modifies its meaning: it is recognized not as a right, but as a vector of risk, an operator of "decentering." It does not attribute to the subject the right to know an object, but to the object the power (to be constructed) to put the subject to the test.

This is thus the abstract definition of the singularity of the modern scientific practices I will propose: if it is no longer a question of vanquishing the power of fiction, *it is always a question of putting it to the test*, of subjecting the reasons we invent to a third party capable of putting them at risk. In other words, it is always a question of inventing practices that will render our opinions vulnerable in relation to something that is irreducible to another opinion. If, as the Sophists said, "man is the measure of all things," it is always a question of inventing practices thanks to which this statement loses it static, relativist character, and enters into a dynamic in which neither man nor thing is the master of measure, where it is the invention of new measures, that is, new relations and new tests, that distribute the respective identities of man and thing.

In order to show that this singularity is constantly being reinvented by the history of the modern sciences, with other givens, and also with other means and other modalities of engagement, I will select first of all a problem being posed today at the heart of the the theoretico-experimental sciences themselves: the apparition of a new type of protagonist that puts in question any possibility of distinction between theory and model.

Mathematical Fictions

The distinction between theory and model, which can seem rather artificial from an epistemological point of view, most often has a very clear meaning from the point of view of the collective practice of the sciences. A model is defined by the absence, at least officially, of any claim to judge: it heralds the absence of the relation of force that would allow it to present itself as the representative of the phenomenon, which can, correlatively, remain explicitly linked to the choice of an author. For a single phenomenon, several models can coexist without any problem, each defined by different variables, each having its zone of privileged validity or its specific advantages.

How should we understand the use of models, in the terms that we have introduced? Models say of themselves that they are fictions, and should be treated as such. But they also constitute a mode of putting fictions to the test, whose stake is not the elimination of rivals, but the following and explicating of consequences. Thus, Samuel Butler's *Erewhon* can be considered as a model. Take the hypothesis of an inversion of our categories concerning those who should be helped and those it would be better to condemn. What does that give us? What will vary or remain invariant in society, or more precisely in Victorian society, as Butler conceived it?

Since the Middle Ages, this regulated and exploratory use of fiction has discovered a privileged instrument in mathematics. Take charity, a "uniformly deformed" magnitude (varying in a linear manner in relation to an extensive variable—in this case, time). What does this definition authorize? What does it allow us to "save," that is, to reproduce as a consequence, among all the statements that we can hold about charity?

It was undoubtedly in order to differentiate itself from this use of mathematics that Galileo took such care to emphasize that his mathematical definition of uniformly accelerated movement was not a fiction due to an author. The phenomenon he invented is capable of silencing counterinterpretations, because it is practically defined in terms of variables that allow it to be both described and controlled: these are the variations by which it responds to changes in the value of these variables, which confirm the legitimacy of the one who represents it. In this sense, the link between mathematical representation and experimental representation is not a deep mystery. Each time a "faithful witness" is created, capable of designating its representative, a representation of the mathematical type is also instituted, putting its testimony onstage, like a function of the variables through the intermediary of which it is interrogated.

The use of mathematics that neither expresses nor gives any power to mathematical representation leads us to another possible history, where mathematics would have established privileged ties with the speculative powers of the imagination, and not with a "theoretical truth" of the world. This history, moreover, is present in our own history, including the history of experimental sciences, for the mathematical imagination is ceaselessly surpassing the possibilities or necessities of the representation of the object. But we have been witnessing, in recent years, the production of a new possibility of history. To some eyes, the use of mathematics as an instrument of fiction could indeed constitute a new future, which would relegate our "Galilean" past and our "Galilean" present to the status of a transitory period whose parenthesis is ready to be closed.

This new putting in perspective is linked to the development of information technologies. In fact, the power of the computer as an instrument of simulation has led to the emergence, among scientists, of what one could call "new Sophists," researchers whose engagement no longer refers to a truth that would always silence fictions, but to the possibility, whatever the phenomenon, of constructing a mathematical fiction that reproduces it.

For example, when Steve Wolfram writes that the universe might be a gigantic computer, we must first of all understand that this universe no longer promises to ground a position of a judge, to consecrate a theory as unifying a diverse field under the unity of a hierarchizing point of view, separating the essential from the anecdotal.[5] In effect, the computer universe establishes a direct relation between phenomenon and simulation, with nothing "beyond" simulation, with no promise of a theory beyond the models. It prefigures the ideal of an ideally versatile matrix able to engender every possible evolution.

Computer simulations not only propose an advent of the fictional use of mathematics, they subvert equally the hierarchy between the purified phenomenon, responding to the ideal intelligibility invented by the experimental representation, and anecdotal complications. Simulation puts what it takes into account on the same plane: "laws" become constraints whose effects have no interest apart from the circumstances that make each simulation a new case. Moreover, what the definition of the "case" itself preserves of the mathematical representation is simply the constraint of a precise, formalizable definition of relations, and not necessarily that of a definition of variables corresponding to the possibility of experimental control.

The art of simulating is that of the screenwriter: to put a *disparate* multiplicity of elements onstage;[6] to define, in a mode which is that of a temporal, narrative "if... then," the way these elements act together; and then to follow the stories that are able to engender this narrative matrix. It is these stories that put the matrix to the test, and make the simulation an experimentation on our statements. They "put them to the act" [*mettre en acte*] without giving us the possibility of intervening, or inflecting the narrative in the direction of what we desire or judge to be plausible. In other words, the characteristic trait of mathematical language, the fact that the statements *engage* [*engagent*], is here extended to the set of descriptions we think of as the "explanation" of a process, and puts them to the test. The explanation, conveyed in the form of a program that will exhibit its consequences, might reveal that it admittedly implies what it was aiming at, but perhaps, in slightly different circumstances, it could imply a very different process—and indeed, if the "dynamic" to which it corresponds is chaotic, this process could be almost anything.

If simulation makes description, explanation, and fiction communicate in a new, experimental mode, and this on all those terrains where authors believe they can propose "reasons" for a history, it poses a specific problem in the theoretico-experimental fields. It is not without reason that the necessity for an "ethics" of simulation has been debated, for the way a program "traffics" in laws, negotiating their scope rather than conveying power, puts into question the mutual mode of engagement between procedure, truth, and reality. The information laboratory is in fact much more rapid, supple, and docile than the material laboratory. There, one can stage phenomena one would not know how to produce in the laboratory, expand certain scales, narrow others, simulate the behavior of a population of a thousand molecules, or subject a crystal endowed with singular flaws to interesting tests. But what does an "experiment" done on an "information" crystal correspond to? Does it produce a fiction or authorize an experimental statement? How should we treat statements of the "experience shows that..." type when it is no longer the question of an event, a conquered link between words and things, but rather a scene that is defined completely in terms of representations?

The "Galileo affair" has bound the experimental sciences against the power of fiction, against the idea that the sole rational vocation for a theory is "to save the phenomena," that is, to simulate them without claiming to penetrate their meaning. One can now conceive of the possibility of a history where the still-open parenthesis will be on the verge of closing, where the power of fiction, affirmed and vanquished by the experimental event, would again become the horizon of scientific practices. This new possibility constitutes, for scientists themselves, a political problem: How should one regulate the relations between the refugees [*ressortissants*] of the two types of laboratory, vectors of divergent modes of engagement? But it is already helping to transform the way certain key stakes in the history of the modern sciences are being formulated, that is, to introduce a form of humor where there now reigns the tragic aesthetic of a reductive science dedicated to leveling all differences.

Very significant, for example, is the contemporary emergence of a field termed *artificial life*. To create artificial life was the dream of the experimenter, the demonstration of the power conquered by humanity on its own conditions of engendering. Now, this field today brings together a crowd of disparate scientists, all those who manage to capture and reproduce a trait of living beings, thanks to recent techniques (robotics, computer simulation). It is no longer a question of reducing but of making proliferate and, correlatively, the alliances no longer rise "to the summit": no discipline is king any longer, the promised site where life will become the object of science. Roboticians and simulators are passionately interested in what ethologists know about such and such behavioral trait, characteristic of a given species, in such and such conditions. Artifice brings into existence [*faire exister*], and to do this it needs a detailed description of what it challenges, but it does not try to make a demonstration. However, it puts to the test the simplistic fictions that underpinned the great putting into perspective of a life whose secret could be revealed, and the putting to the test of the relations between explanation and delegation: "If truly 'to do this, you only have to do that . . . ,' construct for me what this thing you believe you have explained 'will do' through its own activity."

That the sciences of simulation can take the side of diversity, and not the reduction to the same, is not in itself a guarantee of innocuousness. Robots, even if they no longer respond to a vocation of the reproduction of life but the invention of means of delegating to a machinic apparatus one of its traits, have not for all that become gentle and quiet. Indeed, the novelty is rather that, here, the theoretico-experimental enterprise is confronted with other practices, inventive and risky, which by their very existence put in question the power of truth that defines this enterprise. It is not a question of renouncing the distinction between the "artifact" and the "fact created so as to demonstrate," but of becoming interested in something else, in the artifact as such, which itself is also capable of making the difference between human fictions with regard to their possibilities of explaining. Because they use the latest techniques, it is difficult to judge these sciences in terms of lack, obstacle, or immaturity. In fact, through the alliances they create with on-site field specialists, who alone are capable of suggesting which singular traits interest them, they already subvert the order of the disciplines. In particular, they can rely on the impassioned putting into question of the theoretico-experimental model to which, in the name of the field sciences, Stephen J. Gould devoted himself in *Wonderful Life*.[7]

Darwin's Heirs

For several years, Stephen J. Gould published works whose titles—*The Panda's Paw, The Flamingo's Smile, Hen's Teeth and Horse's Toes*—constitute in themselves so many manifestos for the singular novelty of the evolutionary biology descended from Darwin.[8] Novelty in relation to two distinct traditions: on the one hand, that of theoretico-experimental sciences, and on the other, the technico-social conception of living beings, dominant at least since Aristotle.

Judged from the theoretico-technical model, one can wonder if Darwinian biology is indeed a science. American creationists are not mistaken in attacking evolution, and no longer astronomy, like the church of Galileo's era. What "theory" can the Darwinians present to their credit, a theory that would substantiate their power to judge, or differentiate the essential from the anecdotal in an episode of evolution? Are not the great, apparently explanatory concepts—adaptation, survival of the fittest, and so on—revealed a priori to be empty of explanatory power: simple words coming to comment on a history after it has been constituted?

Judged from traditional questions aroused by the difference between the living and the nonliving, the Darwinian response also appears weak. That the criticisms have not taken up the problem of the eye: How can a contingent process, like the one Darwin invokes, produce an apparatus like the eye, if one knows that the slightest defect makes the organ lose all its utility? The eye par excellence represents the "technico-social" conception of the living. It calls for its definition like an instrument, a means in view to an end. The eye is made for seeing. It calls for a conception of the living that would involve the ideal of a society governed by a harmonious division of labor. Each organ, like the eye, does what it has to do for the greater good of the organism, and the latter thus gives its final intelligibility to its parts. How can one not call on a finalizing form of power in order to give an account of this harmony?

Among the heirs of Darwin, there are biologists who accept this challenge as such. They are what one calls neo-Darwinians, who give Darwinian selection such an exhaustive power that it takes the place of the great Engineer, who would have planned the organism in view of his well-conceived interests. Whatever the trait of a given living being, its raison d'être is selection, acting in the midst of a proliferating variety of mutants. Gould has termed this form of Darwinism "Panglossian Adaptionism." "Everything is for the best in the best of all possible worlds," repeats Dr. Pangloss to Candide. Every trait of the living being must be, or must have been, useful, say the neo-Darwinians, because it is its utility that explains its selection.[9]

The critique of the "adaptationist paradigm" is not made in the name of another paradigm, but rather constitutes evolutionary science's adieu to the ambition of judging in accordance with a paradigm. For this ambition was the basis of the power accorded to selection: if it is *the sole* authority that can legitimately give meaning to what is, it justifies the elimination, as a fake [*faux-semblant*] of anything that seems incompatible with the type of temporality invented by Darwin. Darwin's major innovation was undoubtedly the invention of the history of living beings as a *slow* history, adrift, he said, in the sense that it lacks a motor that would have constituted an intrinsic capacity of adaptation proper to life, or the inheritance of acquired traits proposed by Lamarck. It was in the name of this slowness, of the continuous and infinitely progressive action of selection, that Darwin had disqualified, as deceptive, the data of paleontology, which seemed to bear witness to "sudden" mutations (in the scale of geological time). Gould's and Niles Eldredge's theory of punctuated equilibrium has put this judgment in question, and implies that paleontology can become the source of problems, rather than being made dependent on the "adaptationist" narrative. Correlatively, the thesis according to which the massive extinctions would mark the history of living beings calls into question any adaptationist moral: gone are the monotonous and poor histories, whose morals fit so well with our natural judgments. No, the mammals did not vanquish the dinosaurs because the latter were too large, too stupid, at an impasse of evolution, whereas the mammals, which led to ourselves, were already manifesting the superiority that we hold in such high regard.

If selection is not all-powerful, if it does not permit the construction of a point of view from which all cases would come back to the same, and would have the same adaptationist moral, the biologist loses the power to judge and must learn to recount. We here enter into a problematic proper to the field sciences, which distinguishes them from laboratory sciences. One finds at work in the "terrain"—in the depths of the ocean, in the museums where collected fossils are examined, in the forests where samples are harvested—as many sophisticated instruments as there are in an experimental laboratory, as much invention as the meaning of a measure. But one does not find experimental apparatuses in the Galilean sense, giving the scientist the power to stage his own question, that is to say, to purify a phenomenon and give it the power to bear witness to this subject; the instruments of the naturalist, or the field scientist, give him the possibility to collect *indices* that will guide him in his attempt to reconstitute a *concrete* situation, to identify relations, not to represent a phenomenon like a function furnished with its independent variables.[10] Of course, the index, just like the experimental witness, cannot be defined as neutral, independent of the interests and anticipations of an author. But the author, here, knows that his terrain will not make him a judge. No terrain is valid for everyone, no one can authorize the "facts" in the experimental sense of the term. What one terrain allows us to affirm, *another terrain can contradict,* without one of the witnesses being false, or without preventing the two situations from being judged as intrinsically different. Other circumstances have played a role. All the witness of the power of Darwinian selection cannot silence these other witnesses that put in doubt the generality of its explanatory power. The evolutionary biologist no longer knows a priori how the selection plays out in each case nor, above all, what is due to the selection.

Stephen J. Gould's *Wonderful Life* can, on more than one account, be compared with Galileo's *Dialogue.* Here, the power being challenged is not Rome, but the model of the theoretico-experimental sciences. The science of evolution learns to affirms its singularity as a *historical science* faced with experimenters who, whenever there is no "production of facts," can only see an activity of the "stamp-collecting" type.

Today, Darwinian narratives no longer have the moralizing monotony that was dedicated to the triumph of the "best." They make ever more heterogeneous elements intervene, elements that ceaselessly complicate and singularize the narrated intrigue. The living beings are no longer "objects of Darwinian representation," judged in the name of categories separating the essential from the anecdotal. The "concepts" of adaptation and survival of the fittest no longer have the power to make the scientist capable of anticipating the way they will be applied in a given situation. In Darwinian histories, a cause in itself no longer has the general power to cause; each is taken up in a history, and it is from this history that it gains its identity as a cause. Each witness, each group of living beings, is now envisioned as having to recount a singular and local history. Scientists here are not judges, but inquirers, and the fictions they propose take on the style of detective novels, implying ever more unexpected intrigues. Darwinian narrators work together, but in the mode of authors whose intrigues spur each other on. They learn from each other the possibility of making ever more disparate causes intervene, and a mistrust of any cause that carries with it the claim to determine how it causes. In short, mistrust of anything that is identified as a trap: the diverse modes of assimilating of history to progress. In *Wonderful Life*, Simplicio's "role" is played by "our habits of thought," which always tend to define what happened as what had to happen.

The singularity through which I have proposed to define the modern sciences, to invent ways of problematizing the power of fiction and putting it at risk, is thus reinvented here, with other givens. Whereas the experimental apparatus instituted an engagement that could be placed under the sign of the "power to judge," that of the "Darwinian biologist" is inscribed in a strategy of decentering and "demoralization": the aim of the undertaking is to actively allow reality to put our fictions to the test, but it only receives the means to intervene and make the difference in an activity of the "demoralization" of history.

Demoralizing History

We must here understand the term *moral* in the sense in which a moral explanation seeks a cause that would be "worthy" of explaining, that carries in itself the justification of its effect: "better adapted," "more fit"...The moral is always inscribed, from then on, in a perspective of progress, and tends, most often, to put humanity at the center of history. How can one avoid the temptation to judge that between the dinosaurs and the mammals contemporaneous with them there had to exist a difference worthy of explaining the disappearance of the dinosaurs, and the history that led from them to us? Reality, in the Darwinian sense, intervenes to the degree that, even though it is a question of comprehending the history that led to us, this history interests us *in something other than what led to us.*

And in fact, "evolutionists" cannot always recount to us how an eye was created, but they have succeeded in "making history" with living beings, in a way that reinvents the view we have of them. Darwinian efficaciousness lies in the possibility of being interested, as the titles of Gould's collections emphasize, in "strange" traits, in the strange things of nature. The eye will come later, once we are able to liberate it from its image as an instrument to an end, and to comprehend it in terms of even more bizzare histories. As long as we are unable to see the eye as a product of a history, we leave the eye to the side and interest ourselves in the panda's paw, the smile of the flamingo, the migration of tortoises — everything we do not see as long as we think of life in terms of ends. Truth, reality, and procedure are mutually engaged in an operation that creates narrations where we used to understand through judgment.[11]

The procedure of narrating, like the procedure of the experimenter, is a *risky* procedure, subject to the ever-present possibility of creating an artifact. The specific risk of the narrator is tied to the proliferation of indices, which can, as one knows, nourish the power of fiction as well as constrain it. Umberto Eco has created the mythology of this new type of artifact, from *The Name of the Rose*, in which the inquiry of the criminal is guided by pseudo-indices, the correlation between the circumstances of the first crimes and the unfolding of the Apocalypse, to *Foucault's Pendulum*, in which a simple list of deliveries to be made brings into existence the secret society whose existence it seems to reveal.

And the problem posed by the incertitude of the indices is doubled by the one posed by the unstable character, sensitive to the slightest quantitative variation of the simulations models. This is the new horizon of risk opened today by those scientists whom one could call "historians of the Earth," and that marvelously illustrate contemporary controversies with regard to the "greenhouse."

The history of the Earth is now placed under the sign of theatricalization [*scénarisation*], and no longer of judgment, and this novelty is translated by the appearance of scientists provoked by an engagement of a new type, much debated today, for it seems to lead them to intervene in histories that scientists "should not look at." At the beginning of this very interesting history, there was the putting into relation, proposed in 1979 by Luis Alvarez, a physician and geologist, and his son Walter, of a piece of evidence, a narrow layer of iridium distributed in a remarkably homogeneous manner in geological layers corresponding to the end of the Cretaceous era, and a "microfact," the apparently sudden extinction in the same era of between 65 and 70 percent of all living species, including the dinosaurs.[12] Did a giant meteorite really collide with the Earth at this time? Could the collision have unleashed a transformation of meteorological regimes on a planetary scale? Could this transformation have provoked the extinction of the concerned species? In essence, the scenario imagined by Alvarez is interdisciplinary, because it calls for a narrative that integrates the solar flux, climatic variations, meteorological regimes, the behavior of clouds of powder, research on craters, statistics on extinctions, paleontological excavations, and so on. It also constitutes a privileged field open to computer simulation, in the sense that, as we have seen, simulation is naturally interdisciplinary, integrating the play of disparate actors.[13] But it has also been an occasion for a scientific collectivity to recognize the singularity of its practice, and the possibility of new links between human histories and the histories of the processes put into play by the sciences. And this, first of all, from an unexpected question: Could not the simulations produced on the subject of Alvarez's hypothesis (re)become pertinent in the case of nuclear war?

The "nuclear winter" issue, which began in 1983, brought together biologists, meteorologists, and model-producing mathematicians (a regime of interdiscipinary functioning), beyond the cleavages of the cold war (model-producers of all countries, unite!), and it spread disarray among politicians and military men. The threat of nuclear war does not here constitute a "cause" that would have in itself the power to explain the way it has affected scientists (others before them had protested and banded together). Those it united around the theme of a "nuclear winter" were not first of all moral or responsible citizens, but scientists aroused by an event that was "produced" by the encounter between a new possibility of science and the discovery of the unforeseen threat contained in a possibility of history. And the effects of this event have, in the United States, gone beyond the usual "psychosocial" frameworks expected of antinuclear protests. The layer of iridium and the dinosaur fossils, the atmospheric regime and the multiple consequences of climatic variations, have become witnesses of possible histories for a new collective, derailing the calculus of strategies, throwing the Pentagon into a panic, and complicating CIA contacts with the East with regard to modelings, simple speculative modelings (no military secrets, which would have allowed the contacts to be blocked).

It is as scientists that, today, those who try to model the "greenhouse effect," the consequences of deforestation, the effects of pollution, are engaged in, and contribute to, disrupting the politico-economic calculuses. But the "new data" that this new "contingent process" "invents" also sustains new situations of controversy. Scientists, here, are no longer those who bring stable "proofs" but uncertainties.

Irreducible uncertainty is the mark of the field sciences. It does not stem from an inferiority, but from a modification in the relations between "subject" and "object," between those who pose questions and those who respond to them. Correlatively, with regard to the field sciences, it is difficult to speak of "discovery," and the passion to "make exist" then takes on a completely different meaning. No one in fact doubts that the "terrain" exists, that it preexists the one who describes it. Even if it could be said to be invented by numerous procedures that encode and decipher it, it preexists its deciphering in the sense that it presupposes a stability that makes it capable of accommodating interdisciplinary practices. It preexists insofar as these practices presuppose that it is "in principle" capable of making them agree. But furthermore, this preexistence forbids the mobilization, as we have described it. The "artificial" character of the experimental mode of existence can be created, and if this process of creation, as we have seen, makes the theoretico-experimental sciences vulnerable to power, it also confers on the experimental reference a "heavier" existence than that of the terrain.[14] In effect, the terrain does not authorize its representatives to make it exist other than where it is. Nor does it authorize them to prove the stability of the relations that allow it to be described, with regard to change in circumstances or the intrusion of a new element. The dynamic of "bringing into existence," and that of the proof, are no longer an affair of power, but an affair of a process that one must *follow*. The time of the proof, which, in the laboratory, belongs to the sole scientific temporality, is in effect associated with the time of diagnostic processes, time that, eventually, will transform a certain index into a quantifiable, but perhaps irreversible, process. In this sense, field scientists are indeed more annoying than the allies who are interested in power, because they are interested in precisely what power, when it addresses itself to the theoretico-experimental sciences, makes one forget "in the name of science."

"What Does He Want from Me?"

The practice of the theoretico-experimental sciences passes through the event-invention of means for making a phenomenon bear witness, and this invention always implies a systematic putting in variation: it is when it is re-created in the laboratory as a *function* subject to *variables* that a phenomenon becomes capable of designating its legitimate representative. Such a putting in variation is absent in the practices of the field sciences, where each situation can designate its pertinent variables, here and now, without for all that giving the scientist the power to dominate the variety of cases. This variety as such then constitutes the putting to the test of our fictions. But the invention of practices is addressed to beings whose mode of existence *bears witness in itself to the power of fiction*, implies, as we will see, a third type of variation. This time, variation affects scientists themselves, as "modern," according to Bruno Latour's terminology, that is, as seeking to oppose truth and fiction.

Of the Earth, the present subject of our scenarios, we can presuppose a single thing: it doesn't care about the questions we ask about it. What we will call a "catastrophe" will be, for it, a contingency. Microbes will survive, as well as insects, whatever we let loose. In other words, it is only because of the global ecological transformations we can provoke, which are potentially capable of putting in question the regimes of terrestrial existence we depend on, that we can invoke the Earth as having been put in play by our histories. From the viewpoint of the long history of the Earth itself, this will be one more "contingent event" in a long series. This aesthetic of contingency at the same time defines the force, the intrinsic limits, and the style of science practiced by the historians of the Earth, just like the historians of human histories, who address themselves to these factors as "taking part of the past." This style has an analogue in the styles of fiction: the characteristic of the classical detective novel, for example, is that the difference between detective and suspects is stable. The crime, if it took place, took place *before* the detective's intervention. The rule of the genre, in historical narratives, is of the same type: the traits that interest them have a stable identity in relation to the type of intervention that lets them be studied.

The situation of scientific authors, however, is completely different. What they are concerned with — rats, baboons, or humans — are able to "be interested" in the questions asked them, that is, *to interpret from their own point of view* the meaning of the apparatus interrogating them, that is, again, to make themselves exist in a mode that actively integrates the question. The situation is completely different when the history through which the interrogator seeks to become an author *also makes history* for the one being interrogated, that is, when the conditions for *the production of knowledge* of the first are equally, and inevitably, conditions for *the production of the existence* of the second.

If nuclear war can serve as an emblem of the new engagement created by the histories of the Earth, the adventure of apes that "speak" — Sarah, Washoe, Lucy, and so many others—can serve as an emblem for the problem created by the inseparable character of the productions of knowledge and existence. Can chimpanzees learn to speak? The responses brought to this question have aroused and still sustain numerous controversies, which, moreover, have enriched the description we give of human language and its apprenticeship. Likewise for the type of "consciousness" we can attribute to chimpanzees, gorillas, and ourselves. But the price of this production of knowledge is the production of new beings, beings whose potential competencies we "reveal" by plunging them into an intensely human universe, where the questions that create meaning for us take on meaning for them. The "psychoprimatologists" have problems other animal psychologists do not have. They cannot get rid of their experimental material after use — returning them to their natural environment, for example, or to a zoo — for they are hybrid beings, literally "placed in the human world," for whom they feel as responsible as parents do for their children. These links, created in the name of the production of knowledge, link and bind humans to the unforeseen beings they have *brought into existence.*

When the posed question interests the one who asks it as well as the one being asked, although in different modes, the power of fiction also intervenes twice: on the side of scientists, who have to invent a practice that puts their fictions to the test, and on the side of what is not exactly a terrain (although one speaks of terrain in the social sciences),[15] for the question "What does (this scientist) want from me?" is a prodigious resource for speculation and self-production, whether the latter is expressible verbally or is translated by conjectural or perplexing behaviors. The notion of witness becomes ambiguous, scarcely dissociable from the arti-fact (in the negative sense). Correlatively, "bringing into existence" and "proving the existence of" cease to be correlated. It is here that the scientist encounters, on his own terrain, the "charlatan," someone who, for example, mistakes a cure for a proof. It is here that the scientist, so as not to resemble a charlatan, can be tempted to disqualify any question related to the difference between a physicochemical body and a living being (this is only a placebo . . .).

Once again, the question concerning the relation between "sub-ject" and "object" is thereby modified. Someone who, like Stanley Milgram, maintains the usual role of the subject, who takes the initiative of posing questions, which those he deals with will have to respond to, in one way or another, can, in the name of science, "bring into existence" the executioner he believed he had "discerned." The new test, to which the "subject" is submitted, is to have to deal with beings who are able to obey him, to try to satisfy him, to agree, in the name of science, to answer uninteresting questions as if they were pertinent, and indeed, to let themselves be persuaded that they are pertinent because the scientist "knows better"; in any case, with beings that *no means can render indifferent to the fact that they are interrogated*. A being who is interrogated, who is put in the service of knowledge, cannot let itself be put in question without, uncontrollably, the scientific question also taking on a meaning for it. The "object" here looks at, listens to, and interprets the "subject."

It is hardly surprising that, in most cases, the relation between the production of knowledge and the production of existence today looks like an obstacle to scientificity, from experimental psychology to pedagogy, from sociology to medicine, from animal ethology to social psychology. Even psychoanalysis, whose field seems to be defined by this relation, can be described by the ambition to bypass its implications, for it is indeed what permits the staging of the Freudian uncon-scious. Throughout all its theoretical mutations, it has always remained capable of guaranteeing the difference between what would be revealed by the simplest sugges-tion, that is, from the illegitimate power of fiction, and what would be the "truth," irreducible to this fiction.[16] This is because what is here put in question is the ideal that the modern sciences have reconquered, despite Étienne Tempier's verdict, and carried to a new intensity—the ideal of a truth capable of being opposed to fiction, which is also the ideal of a "reality" capable of putting the power of fiction to the test.

Up to now, the right of science to destroy or mutilate what is unable to resist it has been posed above all in ethical terms. Thus, we have no right to subject humans, or indeed living beings, to just any type of interrogation whatever, in the name of science. But questions and procedures that wound dignity or damage health are not the only problematic ones. Every scientific question, because it is a vector of becoming, involves a responsibility. "Who are you to be asking me this question?" "Who am I to be asking you this question?" These are the interrogations that the scientist, who knows the irreducible link between the production of knowledge and the production of existence, cannot escape.

Rather than a strictly ethical question, this is much more a question of what Félix Guattari has called a "new aesthetic paradigm,"[17] where *aesthetic* designates first of all a production of existence that concerns one's *capacity to feel*: the capacity to be affected by the world, not in a mode of subjected interaction, but rather in a double creation of meaning, of oneself and the world.[18]

A contingent recommencement "with other givens"? If we remember the problem, reinvigorated by Marx, concerning the relationship between "science" and "engaged action," as well as Freud's obsession to establish a strict distinction between psychoanalysis and suggestion, we could say that the recommencement has already commenced.[19] The difficulty encountered head-on shows the pertinence of the question. One way of formulating the challenge we have inherited would then be to become capable, one day, of reading Marx or Freud just as biologists can today read Darwin. With tenderness.

In fact, it is profoundly significant that the risks of such a recommencement are explored most explicitly in ethnopsychoanalysis, as defined by Tobie Nathan:[20] to manage to think the Djinns, the spirits of the ancestors, or the most exotic divinities as neither "truly true" nor fictive, but *in the same manner as the Freudian unconscious*, as a constituent part of a psychotherapeutic apparatus; and to manage to avoid thinking the open ensemble of these apparatuses, and the cultural spaces they presuppose and institute, under the sign of a more or less ironical relativity (anything goes), so as to recognize it as the very terrain where the knowledge of what we call "psychism" is constructed—that is, above all, the terrain where those who would have to be capable of experimenting with it, and transmitting its practice, are constructed.[21]

This is something that can wound our Western desire to do science, to create a theory that allows us to distinguish the rations from the irrational. However, what is also at play here is the possibility of a practice that, while putting our fictions to the test, *as required by the singularity of the modern sciences*, creates a position of humor, in which Western culture, as it produces science, submits itself to the most demanding test: the test that reinvents the West as one culture among others. For the fiction that is put to the test by the question of beings capable of transforming every theory into a fiction, and certain fictions into vectors of becoming, is nothing other than our belief in the power of truth, if it is truly true, to denounce fiction.

It is useless to say that scientists engaged in the invention of practices of this kind would no longer constitute mere annoyances, bearers of uncertainty, but would become true traitors, able to follow, in the name of science, the effects of every division, great or small, that allows us to classify, evaluate, judge, identify, silence, and make speak. It is hardly surprising that, today, those who must be termed "maximally objective," according to the criterion proposed by Sandra Harding—inclusion in the scientific practice of putting to the test the relation between the "social experience" of scientists and the "types of cognitive structures" their enterprise privileges—are resolutely marginal.

N I N E

Becomings

How to Resist?

"THE FEELING of shame," Deleuze and Guattari have written, "is one of philosophy's most powerful motifs."[1] But "books of philosophy and works of art . . . have resistance in common — their resistance to death, to servitude, to the intolerable, to shame, and to the present."[2] I am not sure I have been able to write a book of philosophy, but in any case I have tried to work at an experimentation of concepts, which permits us to resist the present, and to appeal to a future in the mirror of which our present and our past "are strangely deformed."[3]

It is not easy to resist without reference to a past we would like to regret, and all the less so insofar as what we are resisting designates this past as outdated, and the future as a promise already disqualifying the present.

Nonetheless, despite the shame that should inspire what has been committed in the name of progress thus defined, do we have the means to take as a reference the regret of a past "that does not progress"? Do we have the means to ourselves pass from a reference to progress?

Whether we speak of science or society, progress is the dominant image; it is what allows us to structure history, to separate the essential from the anec-dotal, to make narration or signification communicate. For us, progress truly con-stitutes both a measure of the march of time and the identifying mark that authorizes the person who speaks to judge. It also authorizes us to simplify the narratives, because progress allows us to select, in a given situation, which narratives are illusory and which are truthful. Progress selects between what is worth conserving and amplifying and what can, with some transitory pains, be relegated to the past. It thus authorizes us to treat the problems of the present in two radically different ways, depending on whether these problems herald the future or represent a past destined to be superseded.

The image of progress is a powerful one. Even the denunciations of episodes once judged by many to be "progressive"—such as colonization, the development of technologies, ideological mobilization—are made in the name of progress, for it is difficult to avoid summary phrases of the type, "We used to believe that . . . but today we know that. . . ." Even the denunciation of Western arrogance, which is thought to be intrinsically different from that of other cultures, does not annul the difference: we are the ones who are in movement, who made others submit to us, but who have now become capable of recognizing our excesses. No "relativist" conclusion can make us forget that, whether as rationalists or relativists, it is always we ourselves who are speaking.

"Before, we did not know what we believed; today, we know that we can no longer believe." The turning point that signals progress is always there. And it still subsists through the ruses and syntactical contortions of the "postmoderns," who glory in no longer believing and dedicate their irony to the description of those who "still believe," little academic games reserved for people of independent means who are the beneficiaries of what they are supposed to no longer believe in. In fact, I do not think we can renounce the reference to progress, for we no longer have any choice: once the question is posed for us, we are defined as the heirs of this reference, free to redefine it, perhaps, but not to annul it. And the interest that "we know that we can no longer believe" is then the problem this phrase sets forth. Knowing we can no longer believe does not mean "ceasing to believe" or ridding ourselves of our heritage—it was a misunderstanding or an error, neither seen nor known—but rather learning to prolong that heritage differently.

The question is thus one of knowing what "we no longer believe" can make us capable of: what sensibilities, what risks, what becomings can it engage us in? Can we confer a positive meaning to "what we no longer believe"? Can we transform the shame of what our beliefs have permitted into a capacity to problematize and invent—that is, to resist?

In a page with prophetic resonances, Bruno Latour evokes the "Parliament of Things." Inside the Parliament, there are no more naked truths, but there are no more naked citizens either. The mediators have the whole space to themselves. The Enlightenment has a dwelling place at last. Natures are present, but with their representatives, scientists who speak in their name. Societies are present, but with the objects that have been serving as their ballast from time immemorial. Let one of the representatives talk, for instance, about the ozone hole, another represent the Monsanto chemical industry, a third the workers of the same chemical industry, another the voters of New Hampshire, a fifth the meteorology of the polar regions; let still another speak in the name of the State; what does it matter, so long as they are all talking about the same thing, about a quasi-object they have all created, the object-discourse-nature-society whose new properties astound us all and whose network extends from my refrigerator to the Antarctic by way of chemistry, law, the State, the economy, and satellites.[4]

Does this baroque image of the Parliament of Things (which is here, as one might have guessed, discussing the ozone hole) reveal a reformist or a revolutionary perspective? This is a question my students often ask, and to which there is no response. The great interest of this image is that it provokes an immediately operative "deformation" of the present under the effect of a future whose demands are without limits. Consequently, it puts in paradoxical communication everything that progress, in the classical sense of the term, was suggesting we oppose, the reformism that humanizes and arranges in continuity, and the revolution that denounces and creates ruptures.

We could say that the Parliament of Things in fact celebrates the triumph of scientific practices; for it constitutes the generalized putting to the test of fictions, and first of all the fiction of a general interest in the name of which particular interests should be submitted. But it recognizes these practices to the degree in which they make the representatives proliferate, always more varied and demanding, and not where they affirm a right.

At the heart of the Parliament of Things, the "patron," Jean-Pierre Changeux or Daniel Cohen, would represent the pandorine, the populations of interconnected neurons, the human genome, but they would rub shoulders, in a stable manner, with representatives of mysticism, of the unconscious, of the set of practices they define as fallow terrains, available to their advance. Their ardor would not have been restrained by limits imposed from without, in the name of an authority whose respect, it would have been decided, must be imposed (fiction instituted as taboo). It would have to invent the means of becoming interested in others and

THEORY OUT OF BOUNDS

in making them interested, with no hope of being able to substitute itself for them "in the name of science." The principle of conquest, in which the indigenous is defined a priori from the viewpoint of its availability to submission, would in effect give way to the principle of multiplicity: every new representative *is added to the others*, complicating the problem that brings them together even if it claims to simplify it; and the representative can only bring into existence what it represents if it succeeds in situating what it represents "between" itself and the others, thereby making itself actively interested in the others in order to comprehend how it can make them interested in itself.

If "Boyle," in this fiction, wins out over "Hobbes," if the multiplicity of representatives of particular interests wins out over the Leviathan of a general fictive interest to which the particular would have to be subjected, the price to pay is clear. The work of mediation, which becomes, as Latour writes, the "center" of the double power, both natural and social, will find itself *slowed down*. Speed, as the principle of mobilization, presupposed an available world, whose contours would be deciphered in terms of obstacles to be skirted, reduced, or ignored. If the contours are populated with "colleagues" whose interests and practices can be modified, but whose legitimacy can be contested, then this mode of mobilization becomes counterproductive. The scientists who "leave their laboratories" to defend the public interest of what they represent would know that clichés (progress, suffering, the possibility of acting, objectivity) through which they today separate what counts from what does not count will disqualify them as surely as an experimental artifact. And the "profile" of the scientist would thus be transformed, becoming as different from the profile of the patron or of the current profile of the scientist from the viewpoint of a "normal" science, as the latter is today different from the profile of Professor Sunflower.

The Parliament of Things has the virtue of humor, which alone is capable of resisting without hating, without denouncing what it is opposed to in the name of a higher force. As Latour writes, it is not "revolutionary" because it already exists, in the sense that there exist multiple networks where representatives discuss, negotiate, and mutually interest each other. But neither is it "reformist," because it brings about a passage to the limit: the network is affirmed as a rhizome, without limits, without a principle of exclusion, without the "judgment of God" that designates a difference in level delimiting the external and the internal, or a priori disqualifying a particular interest as "corporatist."[5] And to the degree that it mines the stable soil of a series of evidences, and provokes problems wherever solutions are reigning, it constitutes a "concept," in Deleuze and Guattari's sense of the term, for whom the "the creation of concepts in itself calls for a future form, it calls forth a new earth and a people that do not yet exist."[6]

"We do not lack communication. On the contrary, we have too much of it. We lack creation. We lack resistance to the present."[7] The Parliament of Things does not belong to the future, like a utopia that would have to be realized — it is not "realizable." It belongs to the present as a vector of becoming or an "experience of thought," that is, as a tool of diagnosis, creation, and resistance.

Nomads of the Third World

In a sense, the Parliament of Things is Popperian. It celebrates the emerging dynamic of these "third world" inhabitants, who are recognized by their capacity to sustain problems beyond beliefs, convictions, and plans. Only humans have seats in it, are seated there, but these humans are defined not as free subjects, characterized by their convictions and ambitions, but as *representatives* of a problem that engages and situates them. Only humans have seats in it, but these humans are not united by a dynamic of intersubjectivity. On the contrary, they have to invent links within disparity, they have to bring into existence rhizomatic prolongations that refer not to a general interest stronger than any one of them but to new interests provoked by their coming together. Which is to say that the Parliament of Things imposes a drastic mutation on third world inhabitants, depriving them of any claim to differentiate "objective knowledge" and politics.

For Popper, the typical inhabitant of the third world was the mathematical statement. The theorematic definition of the irrational number appropriates a set of mathematical practices, detaches them from the terrain where they had meaning, and transforms them into consequences authorized by an ideal form, from the viewpoint of which the set of these terrains becomes a homogeneous space. But this definition opens up a new field to mathematics; it provokes a becoming of mathematics and mathematicians that expresses the transformation of the relation of force between problem and convictions. In other words, the Popperian inhabitant of the third world refers to what Deleuze and Guattari have called, in *A Thousand Plateaus*, "royal" science. "Royal science is inseparable from a 'hylomorphic' model implying both a form that organizes matter and a matter prepared for a form."[8]

Royal science does not make the "ambulant" or "nomad" sciences that preceded it disappear. The latter do not link science and power together, they do not destine science to an autonomous development, because they were in solidarity with their terrain of exploration, because their practices were distributed according to the problems provoked by a singularized material, without having the power to assess the difference between what, from singularities, refers to "matter itself" and what refers to the convictions and ambitions of the practitioners (belonging henceforth to the second world). Royal science "mobilizes" the ambulant process. "In the field of interaction of the two sciences, the ambulant sciences confine themselves to *inventing problems* whose solution is tied to a whole set of collective, nonscientic practices but whose *scientific solution* depends, on the contrary, on royal science and the way it has transformed the problem by introducing it into its theorematic apparatus and its organization of work."[9]

Thus, this mobilization is not simply rhetorical. It presupposes the event, the invented-discovered possibility of redefining singularities and the problems they were posing, and this from a double point of view. From a first point of view, these singularities are judged in the name of a "form" that has the power to render them intelligible, to "integrate" them, and thus to confer on them an intrinsic status through which they can be deduced or anticipated. But from a second point of view, these singularities are then judged and disqualified in the sense that they had previously created the terrain of a practice, for the latter, annexed in its principle, is henceforth qualified by the "particular," "accidental," and merely "practical" interests that assure it a certain de facto autonomy. The differentiation between royal science and ambulant science lies elsewhere. Thus chemistry is "ambulant" for the theoretical physicist, who is interested, for example, in the diversity of chemical elements, of which only the hydrogen atom is sufficient, according to him, to make the model intelligible (physics understands that, chemistry learns it).[10] In short, we here find once again the hierarchized landscape of contemporary scientific knowledges, in which connections are described as conquest and reduction, and whose status is "in principle" measured at the level of the judgments that assess the difference between the intelligible "same" and anecdotal and subordinate difference.

To refer the invention of the modern sciences to the order of the event and not of right [*droit*], as I have tried to do, is first of all to insist on the difference between the "matters" that royal science presupposes and whose availability it sometimes creates, and those that the laboratory effectively invents. If the laboratory is the place where the coappropriation of matter and idea is created, where an "objective third party" is invented, capable of imposing on humans the putting in risk of their fictions, it is "royal" only to the degree that the practice of the sciences is governed by mobilization. It is the locus of a very singular operation: the creation of a third party to which one can attribute the power to ratify its own identification. But this power, if the mobilization does not transform it into the power to disqualify itself, can also define the terrain of a practice that comes to be added to the others, and that poses, in itself, the problem of its prolongation, of its possibilities to link up with the others.

The mutation is both nil, because scientists, insofar as they do not mime science, are already ceaselessly posing the problem of prolongation and linkages, and drastic, because prolongation and linkage are most often, today, redefined as a confirmation of the power of one pole and the subordination of the other. Thus the theorem, which "is of the order of reasons," is constantly making one forget the problem, which is "affective, and is inseparable from the metamorphoses, generations, and creations" through which the prolongations and linkages are negotiated.[11] Correlatively, what royal science "brings into existence" is not celebrated as a story, the actualization of a new existant through multiple metamorphoses and the addition of ever-new significations in ever-new milieus. The actualization is reduced to a revelation: atoms, the void, the force of gravity, nucleic acid, and bacteria had in themselves the power to exist "for us" in the mode that science was content to "discover."

Conversely, could one conceive of the third world inhabitants as nomads, as producers and products of "objective" manners, putting power at risk for the fiction of posing problems, but without designating an available world, waiting for its objective reduction? It is not without interest that mathematics itself, which created the first theorematic appropriation, seems, at least for certain mathematicians, to engage in it. Thus, René Thom pleads for a form of "nomadic" mathematics, whose vocation would not be to reduce the multiplicity of sensible phenomena to the unity of a mathematical description that would subject them to the order of resemblance, but to construct the mathematical intelligibility of their qualitative difference. The fall of a leaf, then, would no longer be a very complicated case of a Galilean register, but would have to provoke its own mathematics. One could also cite Benoit Mandelbrot's fractal mathematics. Here as well, to "understand" means to create a language that opens up the possibility of "encountering" different sensible forms, of reproducing them, without for all that subjugating them to a general law that would give them "reasons" and allow them to be manipulated.

However, just as the invention of theorematic mathematics does not foreshadow or explain the invention of the modern sciences, neither are the aesthetic, technological, and practical mutations of contemporary mathematics enough to ensure a "demobilization" of the positive sciences.[12] This is the signification of the Parliament of Things, namely, to recall the primary and above all political character of the problem (in the sense, of course, that politics itself is also reinvented through the *explication* of problems provoked by certain inhabitants of the third world). Because we now know the connivance of mobilized scientists with all forms of power capable of extending the scope of their judgments, and with a *general, war-like, and abject* definition of the truth — only that which has the power to resist the putting to the test is true — new constraints have to condition the legitimacy of interventions "in the name of science"; and first of all, one that declares every strategy aiming to mask a change of milieu or signification, that is, to pass from a problematic of linking to a claim of unification, to be *antidemocratic, that is to say, irrational..* We must here speak of constraint and not limit, for the limit separates two possibles that, without it, would have been said to be equivalent. It imposes a difference. The constraint, for its part, implies invention and risk. Without constraint, the networks of invention-discussion will always stop, or will change nature, wherever interest can be demanded and no longer has to be provoked, where social and political stratification authorizes one to denounce resistance as obscurantist, irrational, lazy, and allows one to demand that the interlocutor "first of all" learn the science that goes with it. If there are no constraints, why would scientists refuse the alliance of powers permitting them to disqualify whatever complicates the history they are seeking to construct, confirming their own rationality and the ineptitude of those who doubt it?

"It's the same thing, only more complicated" was the slogan of mobilized science, that which puts difference and the "most complicated" under the sign of the "not yet," of the future in which the "same" will have triumphed in fact as it proposes to already triumph in principle. "What risks does this situation make our judgments run to, what becoming and what sensibilities does it impose on us?" would be the question organizing the Parliament of Things.[13]

The Production of Expertise

It goes without saying that the theoretico-experimental enterprise no longer has the status of a model here. But the challenge of the Parliament of Things is not limited to welcoming the set of descendants of Galileo, or those of Darwin, or those (in the end invented) of Marx or Freud. For scientists, of course, are not the only legitimate representatives of things. They represent things only to the degree that we have succeeded in inventing questions for their subject, which permit them to put to the test the fictions that concern them. But today, most technological-social innovations affect things in much more varied modes than those anticipated by our questions, and thus create a gap between "things," as they are implicated in it, and their scientific representation.

This gap is not ready to diminish — indeed, on the contrary — for each new question reveals a multiplicity wherever our fictions foresee a reality to their resemblance. This gap implies that every innovation is made on the basis of a risk, and that we are not even sure what an innovation is: the quantitative intensification of already-existing putting into relation — indeed, its maintenance in slightly different circumstances — can retroactively be inscribed under the sign of the new and the unforeseen. This is obviously the case par excellence in the controversies over the environment (the ozone hole, the greenhouse effect, and so on), where one discovers how scientific knowledges are partial, hesitant, incapable of permitting the economy the risk of decision, faced with questions they have not posed but which are imposed on us, faced with situations that do not let themselves be staged in the laboratory because they integrate an ill-defined number of interrelated variables.

No political constraint can suppress this risk. On the other hand, it can be actively taken into account. It is in this sense that Bruno Latour foresees not only scientific representatives, but industrialists, administrators, workers, and citizens, in the Parliament of Things — other sensibilities implying the formulation of other problems than those scientists are prone to take into account. But here again, the perspective produced is that of a challenge; for the political constraint — that every proposition passes through those who are the most qualified to put it at risk — presupposes that the production of public expertise is actively provoked.

To illustrate the meaning of this challenge, I will take the story of the three little pigs and the big bad wolf. Whereas the houses of the first two pigs, made of straw or twigs, constitute only fictive solutions to the necessity of "being protected," and will not resist the effective putting to the test, which will make the big bad wolf intervene "truly," the house of the third little pig, made of brick and cement, "truly holds." It is therefore not a question of giving oneself over to the relativist irony that, by reducing all difference to fiction, encourages us to forget that the wolf is not subject to our fictions, that is, to forget that our practices have to "hold up" when faced with a reality that, like the wolf, effectively puts them to the test. However, before listening to the experts discussing bricks and cement, it is necessary to be able to problematize what the brick-and-cement solution takes as acquired, and what the story of the three little pigs, as a moral story, holds as acquired. Would it not be possible to invent other relationships with the wolf? On what does the definition of the wolf as a menace depend — that is, the definition of the problem as a "problem of protection"?

In the Parliament of Things, the first priority would be to research — indeed, to provoke — representatives who can point out the possible distinction between the destructive wolf and other possible wolves, which would in no way (or at least in a different way) be implicated in other stories. The experts in "protection against destructive wolves" would retort, of course, that these other stories are risky, and indeed impossible. But they would have to recognize rather quickly that they are not qualified to speak of other stories, nor to follow through all its consequences the logic of the story they are advocating. Can the wolf be defined as a punctual menace, or, if we do not learn how to define it otherwise, do we enter into a story in which other wolves, even more threatening, will intervene, in which the bricks and cement will no longer suffice, in which we will be taken up in an endless move toward ever more costly and rigid modes of protection?

It is here that, in a slightly unexpected manner, the demands of the "politics of reason" and those of the city, in a more classical sense, intersect, and it is in this sense that I have been able to employ the double qualifier, used rather infrequently, "antidemocratic, that is, irrational." In fact, as soon as one puts aside the classical division of responsibilities, which gives the sciences and their experts the task of "informing" politics, of telling it "what it is" and deciding what it "must be," one comes face to face with an *inseparability of principle* between the "democratic" quality of the process of political decision and the "rational" quality of the expert controversy that the Parliament of Things symbolizes. This double quality depends on the way in which the production of expertise will be provoked on the part of all those, scientific or not, who are or could be interested in a decision.

It is not a question here of having citizens "vote," but of inventing apparatuses such that the citizens of whom scientific experts speak can be effectively present, in order to pose questions to which their interest makes them sensible, to demand explanations, to posit conditions, to suggest modalities, in short, to participate in the invention. This presupposes that the concerned citizens are themselves also representatives of an authority of the "third world," who have the power to situate and to put at risk their personal opinions and convictions: they themselves must be able to speak for more than one, to represent a collectivity that has made its members capable of bringing to light the interests through which it is defined.

Here again, it is not a question of utopia, but of what already exists. We are aware of the role of homosexual groups in the negotiations of the measures to take faced with the AIDS epidemic. The Dutch, who on more than one point demonstrate the example of the inseparability between democracy and rationality, have known how to encourage the association of drug addicts, the *Junkiebonden*, whose claims, when taken together, complicate the problem of the experts in illegal drugs, and become a part of the invention of a solution. The drug addicts, in order to become capable of "taking a position" on the measures that concern them, become capable of suggesting policies that do not define them solely as victims to protect and "heal," nor as offenders to punish, but that are addressed to themselves as if to "citizens like any others."[14]

In other cases, the production of expertise concerns citizens who are not distinguished by any prior singularity. Thus, in 1976, in Cambridge, Massachusetts, Mayor Alfred Vellucci, learning that experiments in genetic engineering had taken place at Harvard University, alerted the population, and the scientists had to enter into negotiations with a group of citizens chosen by their peers to form the "Cambridge Experimentation Review Board."[15] Contrary to the fears expressed by most of the specialists when faced with this intrusion of noncompetent people, the group soon turned out to be a valuable interlocutor when faced with the scientists it made to appear as witnesses. According to Dan Hayes, its president,

> all of the recommendations [in the final report], including some sophisticated measures overlooked or avoided by NIH [National Institutes of Health] officials and experts, came from members of the citizens' group itself, not from its scientific advisors. Over the course of its work, the group had gained both technical competence and self-confidence. Some members who "couldn't even formulate a question" in the beginning learned not only to ask cogent questions but to pursue unsatisfying responses with a series of follow-up inquiries. A few could sometimes even spot instances where a witness was quoting something out of context.[16]

Noncompetent citizens, when they do not have to "learn" science "as at school," but are put in a situation where they can demand that scientists respond to their questions, make the effort to render the "information" they possess pertinent and usable—in short, to address themselves to them as if to interlocutors on whom their work depends—have thus been capable of taking a position on a very difficult technical problem, namely, the norms of the security of research laboratories on genetic engineering. There is nothing unexpected here, only the power of the context, which qualified or disqualified, anticipates and suggests the impotence and submission, or fitness and the authority to speak. In the becoming-collective of the group of Cambridge citizens, as in so many others, the key point had been that the citizens did not have to knock at the doors of the laboratories, but had the power to make the scientists come to them. They did not have to listen to them like neutral authorities telling them what "is," but were able to interrogate them, as representatives of determinate interests, with regard to what "must be." The network of technical and scientific negotiations has no other limits than those of the sites where scientists are at liberty to "create" their authority (and for reasons that, in most cases, do not depend on the scientists).

The Parliament of Things does not designate the utopia of intersubjectivity, but imposes the challenge of what Félix Guattari has called the "collective production of subjectivity." "The diverse levels of practice not only need not be homogenized, linked together under a transcendent authority, but should engage in a process of *heterogenesis*. Feminists will never be implicated enough in a becoming-woman, and there is no reason to ask immigrants to renounce the cultural traits that cling to their being or their national adherence."[17] This process of heterogenesis obviously need not be confused with the formation of a universe of differentiated "ghettos," closed on a particularity cultivated in a fetishistic manner, or demanded in a mode of ressentiment. This is why it communicates with the challenge of the Parliament of Things, in which each participant "comes to a decision" on a "quasi-object that they have all created," but that only represents, in a legitimate manner, the disparate association of practices through which they have created and that connects them together. It is thus a question of a "Popperian" emergence of modes of subjectivation that, having becoming capable of affirming themselves as a constraint for others and having been recognized as such, also become capable of entering into a process in which the consequences of the becoming in which they are engaged, their manner of posing problems that are tailor made for them, and their adherence to a tradition that singularizes them are all put at risk.

The process of heterogenesis, in this sense, is not at all utopian, because it is already at work in scientific controversies. One could say, in effect, that the participants in such controversies have to be on the lookout for any "transcendent authority" that would constitute them as disciples of someone whose statements they accept, but also on the lookout for the transversal consequences in their own field of what is proposed in another, heterogeneous field. The production of existence, in the scientific sense, as well as the demands of a new use of reason that we have invented (and which, no doubt, we have irreversibly invented), have engaged us in a history in which the process of heterogenesis has found its political inscription. The Parliament of Things translates this new definition of politics.

A Return to the Sophists

The Sophist Protagoras, we are told, held that "man is the measure of all things." The meaning of his statement is indeterminate. Most often, of course, it is taken in a relativist sense, and disqualified in the name of an appeal to the truth that man would have the vocation to understand—whatever meaning would later be given to the term *truth*, from Plato to Heidegger, from Saint Augustine to Lacan. It can also be understood in a dynamic, constructivist sense. In this case, measure and becoming are combined, for the term *measure* does not designate the thing without also designating what becomes capable of measuring it, what the created link with the thing provokes in its ethical, aesthetic, practical, and ethological singularity.

One can follow this question in ontological terms, for there is no reason for the term *measure* to remain strictly solidary with human practices. The measure expresses a link that is not to be confused with an "interaction," a link that confers two distinct roles on its two poles, which distributes them into a (quasi) subject and (quasi) object. No more than an automobile is measured by the person it runs over, a storm is not measured by the trees it knocks down. But one could perhaps say that the Sun is "measured" by plants, because their being invents itself by designating the Sun as its source of life. Is this not what we confirm when we measure the well-determined wavelengths of solar light absorbed by vegetables, or characterize the relation between germination and the diurnal period? But this is another story, which should not make us forget the singularity of the one I have tried to characterize here, the relation between measure and politics.[18]

"Not all measures are equivalent" is a general statement about what differentiates measure from other types of relation, and we could formulate a distinct version of it in every field where the term *measure* can take on a meaning. Its properly political formulation makes the problem explicit: it is then a question of constructing the criteria of a legitimate measure, that is, a measure that allows us to decide the mode by which we designate the one who, legitimately, will be able to speak for more than one. It is perhaps because humans, as opposed to Shirley Strum's baboons, have constructed forms of legitimacy that are more stable than the flux of interindividual relations—which are ceaselessly confirmed, maintained, tested, or challenged—that they have been able to thematize this problem in a secular register (the Greek heritage). And to establish, correlatively, a distinction between "politics" and "opinion," the first creative, in one way or another, of an authority that designates the second as generally irresponsible, mobile, and inconstant.

According to the thesis that runs through this book, we are under the influence of the invention of a different way of doing politics, one that integrates what the city separated: human affairs (*praxis*) and the management-production of things (*technē*). The event, which we have inherited, is that the invention of a new practice of measuring things by humans, centered on the difference between "fact" and "fiction," has created "another way" of doing politics, that is, another principle of distinction between legitimate representation and opinion, and a new type of actor, who is used to putting the claimants to this distinction to the test. This event is not an advent; a general practice of differentiation between the measures humans can propose to things was not born with the invention of laboratories. One might imagine that, in a human world where the set of practical and conceptual measures that ties us to things had not already been destabilized, where all our knowledges and practices had not already been placed under the sign of fiction, that is, opinion, some balls rolling down Galileo's inclined plane might have constituted merely an interesting "gadget" of little consequence. The "laws of nature," which have, in our world, announced their accessible character, express the fact that, in a new mode, the modern sciences have taken up Plato's old project: to create a relation to the truth in the name of which the Sophists could be chased from the city.

"If Westerners had been content with trading and conquering, looting and dominating, they would not distinguish themselves radically from other tradespeople and conquerors. But no, they invented science, an activity totally distinct from conquest and trade, politics and morality."[19] The author of these lines is saying two things at once. On the one hand, he does not think that science is "a completely distinct activity," and thus he is commenting on the belief that permits us, we other Westerners, to think that we are so different from others. But, on the other hand, he is explaining the very formidable weapon constituted by our very specific form of belief, our belief in science as something "completely distinct," which assures us, in principle, a completely different access to the world and to the truth.

To be sure, all peoples believe themselves to be very different from others, but our belief in ourselves permits us to define others both as interesting—it was we who invented ethnology—and as condemned in advance, in the name of the terrible differentiation, of which we are the vectors, between what is of the order of science and what is of the order of culture, between objectivity and subjective fictions. We have ceaselessly denounced looters and tradespeople who exploit and subject, but we believe we know that the "others," in one way or another, will have to go through an abandonment of their cultural "beliefs," which mix together what we separate.

The perspective I have tried to open up in this book is one in which we would have to become even more "different," that is, one in which we would have to invent, in our own terms, an antidote to the belief that makes us so formidable, the belief that defines truth and fiction in terms of an opposition, in terms of the power that makes the first destroy the second, a belief older than the invention of the modern sciences, but whose invention constituted a "recommencement." For me, this perspective corresponds to the double constraint of the event: it makes a difference between the past and the future, in relation to which every dream of a "backwards return" is a vector of monstrosity; it does not have the power to dictate to its heirs how to take it into account. The event, which has constituted the invention of a new meaning for the sophistic statement "man is the measure of all things," does not have the power to constitute us as the hallucinated heirs of this possibility of measure; it situates us in terms of a requirement and not a destiny.

Contrary to habits of thought we owe to a vaguely Hegelian tradition, I have not sought in a "stronger" reference the possibility of "overcoming-going beyond" [*surmonter-dépasser*] our belief in objective truth. It is not a question of creating the position from which we could judge, but of inventing the means of *civilizing* it, of making it capable of coexisting with what it is not, without considering, overtly or secretly, that it has—or that it would have, in principle, if it were not limiting itself—the power to reduce the heterogeneous to the homogeneous; "one more mode of measure" that is added to the others and creates new possibilities of history, and not the final advent of "*the* mode of measure." To highlight the difference between the perspective I am trying to create and a perspective of self-limitation (a vector of what one could call "paternalism," for a radical difference is hollowed out between the authority that limits itself in order not to destroy the other, and the other that survives by the grace of the first), I have tried to put it under the sign of humor. The humor that would permit us to treat the avatars of our belief in the truth as contingent processes, open to a reinvention with "other givens," it seems to me, is vital for resisting the shame of the present.

Humor is necessary in order to keep ourselves from overestimating the heroism of the challenge. We do not have to invent ourselves as radically different from what we are, for we are already very different from what we believe ourselves to be. Thus, we do not have to take on ourselves the heroic task of establishing links bween the two ways of doing the politics we have invented, the one that, officially, only concerns humans, and the one that, apparently, has nothing to do with politics. These links have always existed, and our belief in objective truth has never been an obstacle to it. Scientists have always known how to speak to politicians, and politicians have quickly learned the multiple and interesting possibilities of alliance with scientists. Thus, it is not a question of establishing links, but of inventing-thematizing them as political. This does not mean, obviously, that the choices that, today, are made "in the name of science" or "in the name of rationality" could, as if by a miracle, be returned to those they concern. That is another story, which our belief in truth and progress has served as an alibi, but one must be a Heideggerian or a denunciator of "technoscience" in order to assimilate to that story the submission of the world to the operative rationality of sciences and technologies.

But humor, the art of a resistance without transcendence,[20] is above all linked with a second meaning of the sophistic statement that "man is the measure of all things." It designates the becoming of the one who *becomes capable of measuring*, that is, who also becomes what the measure of the thing demands of him, that to which the thing *obligates him*. "To be the measure of all things," then, designates the human as a passion, as capable of becoming "affected by all things" in a mode that is not that of contingent interaction, but of the creation of meaning. Where the sophistic statement, understood in a relativist mode, seemed to designate a static right of opinion, the triumph of the power of fiction, we can read a characterization of the human adventure that links together truth and fiction, rooting both in the passion that makes us capable of fiction as much as the putting to the test of our fictions.

This is not a "content" that disqualifies opinion, but a differentiation of the political type between two meanings of the term *passion*. Passion signifies submission when a strategy of differentiation anticipates, suggests—and thereby constitutes—those whom it qualifies as submitted. Nor is it a "content" that qualifies statements we recognize as scientific, but the invention of active passions, which imply, suggest, and anticipate a demand that, up to now, scientists have called "autonomy": the creation of modes of controversy that presuppose a passion shared by their participants, and thus a specific milieu—the laboratory, or the "terrain"—which one does not enter the way one enters a windmill. It is not by denouncing that one can civilize this passion of differentiation, but by welcoming it with humor, that is, by presupposing, anticipating, suggesting that scientists are able to know that their passion changes meaning when they themselves change milieu. Which implies, as we have seen, a political problem—that "milieus" not invented by the sciences are not a priori defined as available, that is, as governed by opinion and awaiting rationality, but are actively recognized as populated by different ways of "measuring": ways of posing problems, of evaluating consequences, of inventing significations. It also demands that, in speaking of the way the sciences invent their "measures," we relate them to the style of passion that defines their specific milieu—the affective problem of a humor of truth.

The first invention of the modern sciences—the invention of the experimental sciences—required a style of passion that made the scientific author a singular hybrid, somewhere between a judge and a poet. The scientist-poet "creates" his object, he "fabricates" a reality that does not exist as such in the world but is rather of the order of fiction. The scientist-judge has to succeed in making one admit that the reality he has fabricated is capable of supporting a faithful witness, that is to say, that his fabrication can claim the title of a simple purification, an elimination of parasites, a practical staging of the categories with which it is legitimate to interrogate the object. The artifact must be recognized as being irreducible to an artifact. From the poet-judge passionately participating in a game of which many recognize the malicious humor—to transform an apparently insignificant detail into a difference that trips up a rival colleague—to the prophet announcing what will be, or what should be, we know that the distance is short, all the shorter insofar as it is the "prophet" who is awaited and anticipated by the public. The humor of theoreticians and experimenters does not have any rights outside the homogeneous network of rival colleagues; it is one of the prices they pay themselves for the regime of mobilization that constitutes their model enterprise.

The passion of the "Darwinian narrators" does not make them either poets, in the sense of "fabricators," or prophets, but it makes them vulnerable to irony, for the "measure" of the histories of the Earth that they learn to recount demands of them an "aesthetic of contingency," an engagement that constrains them to treat as "habits of thought," as sources of moralizing fiction, anything that would lead us to overestimate the question of human becomings. Darwinian histories are populated with innovations whose signification is transformed by circumstances, which create—out of small differences, and without any reason that would be superior to them—the disappearance of some and the success, perhaps momentary, of others. The humor of the Darwinian narrator stems from the way he can express both the contingency and the noncontingent demand that make him exist, and that link him to the human adventure.

Humor does not have to be merely the guardrail of scientific passions. It can be the constitutive condition of these passions. And this will be the case if demands are invented where scientists could become the "measure" of becomings that do not authorize the separation between the production of knowledge and the production of existence. For it is no doubt here that the two meanings of the sophistic statement converge: the one that links together measure and politics, and the one that links together measure and becoming. In both cases, fiction becomes the vector of becoming, and the differentiation between legitimate representation and opinion, the power to vanquish fiction that is attributed to truth, becomes the "habit of thought" we have to learn to put at risk. In both cases, our Western passion for truth would then require itself to separate truth from power, and to link truths to becomings.

Notes

1. The Sciences and Their Interpreters

1. See the anthology *La science telle qu'elle se fait*, ed. Michel Callon and Bruno Latour (Paris: La Découverte, 1991).

2. Primarily at the Centre de sociologie de l'innovation of l'École des mines, directed by Michel Callon. See Michel Callon, ed., *La Science et ses réseaux* (Paris: La Découverte, 1989), and the following works by Bruno Latour: *The Pasteurization of France*, trans. Alan Sheridan and John Law (Cambridge: Harvard University Press, 1988); *Laboratory Life: The Construction of Scientific Facts*, with Steven Woolgar, 2d ed. (Princeton, N.J.: Princeton University Press, 1986); *Science in Action: How to Follow Scientists and Engineers through Society* (Cambridge: Harvard University Press, 1988); and *We Have Never Been Modern*, trans. Catharine Porter (Cambridge: Harvard University Press, 1993).

3. Michael Polanyi, *Personal Knowledge: Towards a Post-Critical Philosophy* (London: Routledge and Kegan Paul, 1958). In *The Structure of Scientific Revolutions* (Chicago: University of Chicago Press, 1958), Kuhn stressed the similarity between Polanyi's descriptions and his own.

4. The acts of this congress have been edited under the title *Science at the Crossroads* (London: Frank Cass, 1971).

5. John D. Bernal, *The Social Function of Science* (London: Routledge and Kegan Paul, 1939).

6. Michael Polanyi, "The Republic of Science: Its Political and Economic Theory" *Minerva: A Review of Science, Learning, and Policy*, 1:1 (1962): 54–73.

7. Ibid., 72.

8. See Sandra Harding, *The Science Question in Feminism* (London: Routledge and Kegan Paul, 1986).

9. Today, many researchers, notably physicians and chemists, say that this is precisely what is happening. Sponsoring financial institutions are no longer interested in what promises "applications." Numerous researchers no longer use their instruments except to obtain "numbers" [*chiffres*] that can be useful to industry. Students laugh when one speaks to them of "fundamental questions." I will not pursue here this theme of the "end of true research," which would require studies in the field. I simply wanted to make note of a rather brutal development that has been taking place in the last few years.

10. Alfred North Whitehead, whose speculative audacity is equaled only by Leibniz's monadology, likewise held that "you may polish up commonsense, you may contradict in detail, you may surprise it. But ultimately your whole task is to satisfy it" (*The Aims of Education and Other Essays* [New York: Free Press, 1985], p. 107).

11. "There is no worse persecutor of a grain of corn than another grain of corn that has completely identified itself with a chicken" (Samuel Butler, *Life and Habit* [reissue; New York: Classic Book Distributors, 1999], p. 137).

12. Denis Diderot, *D'Alembert's Dream*, and the interview that follows. See, for example, the Penguin Classics edition, *Rameau's Nephew and D'Alembert's Dream*, trans. Leonard Tanock (New York: Penguin, 1976).

2. Science and Nonscience

1. Sandra Harding, *The Science Question in Feminism* (Ithaca, N.Y.: Cornell University Press, 1986), pp. 248–49. In this context, we must understand "minority" in Deleuze and Guattari's sense, for whom the minority differs from the majority not quantitatively but qualitatively: "All becoming is minoritarian. Women, regardless of their numbers, are a minority.... They create only by making possible a becoming over which they do not have ownership, into which they themselves must enter, a becoming-woman affecting all of humankind, men and women both" (Gilles Deleuze and Félix Guattari, *A Thousand Plateaus: Capitalism and Schizophrenia*, trans. Brian Massumi [Minneapolis: University of Minnesota Press, 1987], p. 106).

2. Harding, *The Science Question in Feminism*, p. 250.

3. See Léon Chertok and Isabelle Stengers, *A Critique of Psychoanalytic Reason: Hypnosis as a Scientific Problem from Lavoisier to Lacan*, trans. Martha Noel Evans (Stanford, Calif.: Stanford University Press, 1992), in which we present the inquiry led in 1784 (by a commission that included the greatest scientific figures of the period, among them Lavoisier) on the magnetic practices of Mesmer as the inaugural act of this definition of scientific medicine, and study its consequences through the problem of hypnosis and psychotherapy.

4. On this subject, see the "historical" work of Elisabeth Roudinesco as well as Léon Chertok, Isabelle Stengers, and Didier Gille, *Mémoires d'un hérétique* (Paris: La Découverte, 1990), for the role of the "break" or "cut" in the question of the relations between hypnosis and psychoanalysis.

5. Gaston Bachelard, *La Formation de l'esprit scientifique* (1938; Paris: Vrin, 1975), p. 14.

6. Except, of course, the new production of science. In *Mémoires d'un hérétique* (see note 4), we take as an example the argument of the psychoanalyst Octave Mannoni with regard to the question of hypnosis: we have to "wait for the genius" who will make hypnosis an object of science. Insofar as it is an "annoying" phenomenon, without any positive characterization, its interest is not "a cause to defend"; it does not have the title to put in question the categories of those practices that have conquered the power to define their object.

7. Bachelard, *La Formation de l'esprit scientifique*, p. 251.

8. See Ilya Prigogine and Isabelle Stengers, *Entre le temps et l'éternité* (Paris: Fayard, 1988). The reduction of thermodynamic entropy to a dynamic interpretation can hardly be judged as anything other than an "ideological claim," but it lies at the origin of a history independent of which the physics of the twentieth century cannot be recounted. [A revised version of this book has been published in English by Ilya Prigogine as *The End of Certainty: Time, Chaos, and the New Laws of Nature* (New York: Free Press, 1997). — *Trans.*]

9. See Alan Chalmers, *What Is This Thing Called Science? An Assessment of the Nature and Status of Science and Its Methods* (St. Lucia: University of Queensland Press, 1982).

10. See Gerald Holton, "Mach, Einstein, and the Search for Reality," in *Thematic Origins of Scientific Thought: Kepler to Einstein* (Cambridge: Harvard University Press, 1973).

11. See Imre Lakatos, "Falsification and the Methodology of Research Programmes," in *Criticism and the Growth of Knowledge*, ed. Imre Lakatos and Alan Musgrave (Cambridge: Cambridge University Press, 1970). This book can be considered as the point of "achievement" [*achèvement*], in both senses of the term, of the demarcationist tradition. It is the result of a conference held in 1965 in order to bring the positions of Popper and his principal students into dialogue with those of Thomas Kuhn.

12. Pierre Duhem, *The Aim and Structure of Physical Theory*, trans. Philip P. Weiner, foreword by Prince Louis de Broglie (Princeton, N.J.: Princeton University Press, 1954).

13. To unite ethics, aesthetics, and ethology as I have done here is not unrelated to the notion of "existential territory" introduced by Félix Guattari in *Chaosmosis: An Ethico-Aesthetic Paradigm*, trans. Paul Bains (Bloomington: Indiana University Press, 1995).

14. This is what allows Raymond Boudon, in *The Art of Self-Persuasion: The Social Explanation of False Beliefs*, trans. Malcolm Slater (London: Polity Press, 1997), to define the criterion of demarcation as relevant to a "hyperbolic theory," that is, a theory that leads to conclusions whose generality dissimulates its implicit and debatable a prioris. Boudon, for his part, is satisfied with a tranquil ("polythetic") characterization of the sciences, which allows him to welcome as "theories," and indeed as "laws," the set of general statements accepted by the social and economic sciences. The question of the singularity of the sciences—a question I share with Popper—is thereby emptied in favor of an ecumenical vision: in each domain, one could say "we can do better," and good sense would suffice to recognize the multiplicity of significations that reconfigure the terms serving as criteria for this "better": progress, truth, theory, rationality, and so on.

15. Imre Lakatos, "Replies to Critics," in *Boston Studies in Philosophy of Science*, vol. 8 (Dordrecht, The Netherlands: D. Reidel, 1971).

16. Paul K. Feyerabend, *Against Method: Outline of an Anarchistic Theory of Knowledge* (London and New York: Verso, 1975).

17. Bruno Latour, *We Have Never Been Modern*, trans. Catharine Porter (Cambridge: Harvard University Press, 1993), p. 116: "The words 'science,' 'technology,' 'organization,' 'economy,' 'abstraction,' 'formalism,' and 'universality' designate many real effects that we must indeed respect and which we have to account for. But in no case do they designate the causes of these same effects. These words are good nouns, but they make lousy adjectives and terrible adverbs."

18. See the chapter titled "Trivializing Knowledge: Comments on Popper's Excursions into Philosophy" in Paul K. Feyerabend, *Farewell to Reason* (London and New York: Verso, 1987), pp. 162–91.

19. Paul K. Feyerabend, "Notes on Relativism," in *Farewell to Reason*, pp. 19–89; 30.

20. To speak like Luc Ferry, in *The New Ecological Order*, trans. Carol Volk (Chicago: University of Chicago Press, 1995), who constitutes a good example of scientistic humanism.

21. Paul K. Feyerabend, "Farewell to Reason," in *Farewell to Reason*, pp. 280–319; 297.

22. Ibid.

23. Ibid., 303.

3. The Force of History

1. Bernadette Bensaude-Vincent, *Études sur Hélène Metzger*, ed. Gad Freudenthal, in *Corpus* 8–9 (1988), a journal on the corpus of works in philosophy in the French language.

2. For an attempt to take this antagonism into account actively, see Bernadette Bensaude-Vincent and Isabelle Stengers, *A History of Chemistry*, trans. Deborah Van Dam (Cambridge: Harvard University Press, 1997).

3. Let us here cite the fine book by Trevor Pinch, *Confronting Nature: The Sociology of Solar-Neutrino Detection* (Dordrecht, The Netherlands: D. Reidel, 1986), which retraces in a completely passionate manner the construction, by Ray Davis, a pioneering specialist in the detection of neutrinos, of the "neutrino solar" object, in the sense that the latter realizes a new encounter between physical disciplines that have hitherto been disjoint. It happens that the measure of the flow of neutrinos emitted by the Sun did not give the values foreseen by the model implying astrophysics, the science of nuclear reactions, physics of the neutrino. Which was put in

question? For twenty-five years, the question has remained open: the measurement was confirmed, and the *anomaly* is thus recognized. Pinch's book is a good example of historicizing, but it takes *advantage* of the uncertainty of the actors to demonstrate that science is a matter of interpretation. What he does not emphasize, by contrast, is that the interpretive activity of the actors would have been very different—and that the question would undoubtedly not have remained open—if these actors had not been convinced that the anomaly *could* be resolved, that is, that a response will be able to be produced that renders, after one or another modification, the encounter of the disciplines coherent with the measurement. Whoever realizes this "progress" will no doubt receive a Nobel prize, but the study of the same case by a future sociologist will give *less easily* to the latter the power of differentiating his position from that of his actors: "Of course, for the scientists, nature appears as an independent kingdom, existing objectively. But for the sociologist, nature can only be accessible through discursive processes" (pp. 19–20). The scientist will be able to reply: "Certainly, but here again, it has been rendered 'truly' accessible; all the discursive processes are not valid."

4. In biology, this principle of proliferation is *sometimes* pertinent, notably with regard to bacteria. It is this principle that brings into play the procedures of laboratories in which research into a particular mutant stump is done by supposing that it "must indeed" exist in the population and by submitting this population to conditions such that only these mutants survive.

5. Popper in this way justifies the triumph of "internal" history over external history. Every time a partisan of "external" history wants to correlate the position of a scientist participating in a controversy with his cultural, social, and political interests, the internal historian can say that the first raison d'être of the controversy is an objective problem. The way the actors distribute themselves around the problem on which the conflict depends creates the possibility that the interests in conflict can create scientific divergences. See notably the response of Alan Chalmers, in *Science and Its Fabrication* (Minneapolis: University of Minnesota Press, 1990), to Donald MacKenzie's study, "How to Do a Sociology of Statistics . . ." (included in *La science telle qu'elle se fait*, ed. Michel Callon and Bruno Latour. [Paris: La Découverte, 1991]).

6. Other modes of history are pertinent, and notably the one that Daniel Bensaïd, in *Walter Benjamin, sentinelle messianique: À la gauche du possible* (Paris: Plon, 1990), names "historical materialism," in which the historian knows that it is less a question of reconstituting than of remembering and watching, in a present "summoned to take over from exhausted sentinels in the empty desert, for the case in which there appears a Godot in heels" (p.

94). This present, "which is by no means a passage but which remains immobile on the threshold of time . . . is the time of politics. Every event of the past can acquire or refind in it a higher degree of actuality than that which it had at the moment it took place. History that claims to show how things have really happened is animated by a detective [*policière*] conception that constitutes 'the most powerful narcotic of the century'" (p. 68).

7. Thomas Kuhn, "Reflections on My Critics," in *Criticism and the Growth of Knowledge*, ed. Imre Lakatos and Alan Musgrave (Cambridge: Cambridge University Press, 1970), p. 263n.

8. As Margaret Masterman emphasizes in *Criticism and the Growth of Knowledge*, the definition of the paradigm, in *The Structure of Scientific Revolutions*, is rather imprecise (she counts twenty-one distinct meanings). Contrary to what is often claimed, faced with this critique, Kuhn has less modified his notion than he has learned how much he had to specify his notion in order to avoid misunderstandings. In the strict sense, the question of the paradigm is linked to that of the modern sciences. In other words, it excludes the possibility of speaking of an "Aristotelian paradigm of movement."

9. This is a central theme of the description that Ian Hacking gives of experimentation. See his *Representing and Intervening: Introductory Topics in the Philosophy of Science* (Cambridge: Cambridge University Press, 1983).

10. As Kuhn says, in "Reflections on My Critics," incommensurability is neither more nor less dramatic in science than in different natural languages: a translation, though never perfect, is always possible, but it does not simply make a third "neutral" language intervene. Translators speak the two languages, and seek to negotiate the best compromise between the constraints and possibilities that singularize each of them. This implies that apprenticeship to a paradigm, any more than to natural languages, is not intrinsically linguistic.

11. Let us recall that a translation is not a necessary consequence. It merely designates "that which" is the object of a translation, as a necessary condition.

12. In the sense of Umberto Maturana's and Francisco Varela's theory of autopoesis.

4. Irony and Humor

1. The "strong" program was defined by David Bloor in *Knowledge and Social Imagery* (London: Routledge and Kegan Paul, 1976). This program affirms that the totality of scientific practice, including the distinction between truth and error, is the spring of sociological analysis, and that the adhesion to a scientific theory refers to the same type of explication (psychological, social, economic, political, etc.) as a belief. This strong program is associated with the schools of Bath (Harry Collins, Trevor Pinch) and Edinburgh (Barry Barnes, David Bloor).

2. The interest of *Criticism and the Growth of Knowledge* lies in the confrontation of these two "close neighbors."

3. For a conception of the "human sciences" that resolutely blurs the difference I am constructing here, see the different books of the Marxist philosopher Roy Bhaskar, and notably *The Possibility of Naturalism: A Philosophical Critique of the Contemporary Human Sciences* (Brighton, Sussex: Harvester Press, 1979).

4. Let us note the parallel between this questioning of the power to judge and the singularity of the science of living beings as characterized by Popper's "second world." The whole stake of this second world is to indicate that the biologist must *follow* the invention, by the living being, of the meaning that it or its species will give to questions such as "How to reproduce?" "What relation should be retained between fellow creatures, prey and predators?" "What part of individuality should be linked to apprenticeship, and what other part to the repetition of a specific identity?" In this sense, the science of living beings, like that of politics, cannot be reductive, since neither of them can "precede" what they are concerned with by a general definition of what are good variables to take into consideration, and what are negligible anecdotal dimensions. Both are concerned with a set of "beings" that are so many formulations of this problem, definitions of its variables, inventions of its solution.

5. Hannah Arendt, *The Human Condition*, 2d ed. (Chicago: University of Chicago Press, 1998), p. 3, as cited in Barbara Cassin's article "De l'organisme au pique-nique," in *Nos grecs et leurs modernes*, ed. Barbara Cassin (Paris: Seuil, 1992), pp. 114–48. See also Jacques Taminiaux, *The Thracian Maid and the Professional Thinker: Arendt and Heidegger*, trans. and ed. Michael Gendre (Albany: State University of New York Press, 1998), for the debate on Aristotle.

6. Shirley C. Strum, *Almost Human: A Journey into the World of Baboons* (New York: Norton, 1990).

7. Let us respond to a curious development of this difference. Priests of Kataragama, in the south of Sri Lanka, have successfully brought to trial, for insulting believers, an ethnologist who in their eyes was guilty of having described their rite (to suspend, by hooks planted in the back, volunteers who were prepared long in advance and were "miraculously" insensitive to pain) in a mode that denies the presence of God, to which, for them, this insensibility attests. It is necessary to reflect before crying obscurantist scandal.

8. Shirley C. Strum and Bruno Latour, "Redefining the Social Link: From Baboons to Humans," *Social Science Information* 26:4 (1987): 783–802; 797.

9. In "Redefining the Social Link: From Baboons to Humans," Shirley C. Strum and Bruno Latour emphasize that the "handicap" of baboons in relation to us, which also marks the difficulty of the primatologist's job, is the precariousness of links: the latter must be ceaselessly maintained, put to the test, confirmed. The "society" of baboons would be in this sense more complex than our own, where *marks* stabilize the links, stratify the interactions, and thereby simplify the labor of situating individuals relative to one another. In this sense, human individuals are characterized by their "relative" *obedience*, by their submission to marks of authority or legitimacy. But captive primates, no doubt, also live in a stable and marked universe where they become capable of new types of links, notably those that lead us to discuss the question of knowing if they "speak."

10. Steve Woolgar, "Irony in the Social Study of Science," in *Science Observed*, ed. Karin Knorr-Cetina and Michael Mulkay (London: Sage Publications, 1983), pp. 239–66.

11. *Aggadoth du Talmud de Babylone: La source de Jacob*, trans. Arlette Elkaïm-Sartre (Lagrasse: Éditions Verdier, 1982), pp. 887–88.

12. Gilles Deleuze, *Difference and Repetition*, trans. Paul Patton (New York: Columbia University Press, 1994), p. 79.

5. Science under the Sign of the Event

1. Gilles Deleuze and Félix Guattari, *What Is Philosophy?*, trans. Hugh Tomlinson and Graham Burchell (New York: Columbia University Press, 1994), p. 98.

2. Trevor Pinch, *Confronting Nature: The Sociology of Solar-Neutrino Detection* (Dordrecht, The Netherlands: D. Reidel, 1986), p. 18.

3. Galileo Galilei, *Discourse concerning Two New Sciences*, trans. Henry Crew and Alfonso di Salvio (New York: Dover, 1954), p. 154.

4. Ibid., p. 161.

5. Cited in Pierre Duhem, *To Save the Phenomena: An Essay on the Idea of Physical Theory from Plato to Galileo*, trans. Edmund Doland and Chaninah Maschler (Chicago and London: University of Chicago Press, 1969), pp. 110–11.

6. Éric Alliez, *Capital Times: Tales from the Conquest of Time*, trans. Georges Van Den Abbeele (Minneapolis: University of Minnesota Press, 1996).

7. Stephen Hawking, *A Brief History of Time: From the Big Bang to Black Holes* (New York: Bantam Books, 1988).

8. The possibility of saying *both* that the subject is "pathological," that is, that what it has done is explainable, and that it is "free," that is, that it could have *not* done it,

is the solution Kant proposes in the *Critique of Pure Reason* ("Solution of Cosmological Ideas That Derive Their Causes from the Totality of the Events of the World").

9. Galileo Galilei, *Two New Sciences*, trans. Stillman Drake (Madison: University of Wisconsin Press, 1974), p. 90 [132], and then pp. 93–94.

10. Ibid., p. 206.

11. I will not here take up the dispute between Pierre Duhem, Alexandre Koyré, and Stilman Drake on the medieval roots of the Galilean conceptions, and on the way the famous letter of 1604 should be read, in which Galileo announces for the first time that he possesses the mathematical definition of accelerated movement, such that all observed experiments are in agreement, and "is wrong." For all this, see Isabelle Stengers, "Les affaires de Galilée," in *Éléments d'historie des sciences* (Paris: Bordas, 1989), pp. 223–49. See also Stilman Drake, *Galileo at Work: His Scientific Biography* (Chicago: University of Chicago Press, 1978).

12. It must thus be emphasized that, although the *Discourse* follows the *Dialogue*, it relates works that took place *before* the astronomical quarrel with Rome. This is why nothing prevents us from thinking that Galileo the polemicist, who tried to force Rome to bow down before the heliocentric truth, was born in the laboratory, one consequence among others of what I am calling the "Galilean event."

13. The ball had to descend the length of an inclined plane, for if Galileo had let it fall, it would have bounced in place of pursuing in a (more or less) continuous manner its movement on the table.

14. This is what has been staged by Didier Gille and Isabelle Stengers in "Faits et preuves: fallait-il le croire?" in *Les Cahiers de Science et Vie: Les grandes controverses scientifiques*, no. 2, *Galilée: Naissance de la physique* (April 1992): 52–71.

6. Making History

1. This process can, moreover, pose the problem to scientists themselves, when the selection-exclusion is made too radical. This is the case in high-energy physics today, where the selection-exclusion is integrated into the experimental apparatus itself: the informational treatment of the data is guided by the theory that qualifies different events, and retains only those that it judges to be significant. Here, the physicists themselves come to ask "where" their own history has led them— without which, nonetheless, they have the means to proceed otherwise.

2. It is not without interest, however, that *New Scientist* (11 July 1992) published a rather positive critique of a book by the present director of research at the Institute of Parapsychology at Durham, North Carolina (Richard Broughton, *Parapsychology: The Controversial Science* [London: Rider, 1992]), which ends with "only time will tell...." And in its 15 May 1993 issue, the same *New Scientist* dedicated its cover to the question ("Telepathy Takes on the Skeptics"), an article by John McCrone, "Roll Up for the Telepathy Test," and concluded that, in the near future, the ball will perhaps find itself in the camp of the skeptics. A matter to be followed.

3. On this subject, see *Nicomachean Ethics*, as well as the "non-Heideggerian-Platonic" presentation of Jacques Taminiaux in *The Thracian Maid and the Professional Thinker: Arendt and Heidegger*, trans. Michael Gendre (Albany: State University of New York Press, 1998).

4. See Michel Callon, ed., *La Science et ses réseaux* (Paris: La Découverte, 1989).

5. Bruno Latour, "D'où viennent les microbes?" in *Les Cahiers de Science et Vie: Les grandes controverses scientifiques*, no. 4, *Pasteur: La tumultueuse naissance de la biologie moderne* (August 1991): 47.

6. Usually, but not always. If "cold fusion" had fulfilled its promises, it would have resembled the discovery of America. The network of interested allies, ready to take it as the resource and referent of their practice, preexisted with such a force that the consequences of this "discovery" had already begun to be reproduced when the rival colleagues of Martin Fleischman and Stanley Pons announced that, from their point of view, the difference between experimental statement and fiction had not been established. Moreover, the active interest of lawyers, concerned with the question of patents, or the interested reference to their demands, gave a rather original feel to the controversy. Here, the interdiction to "enter the laboratory like a windmill" was not addressed to non-competents, but to competent colleagues, who could have later claimed rights to the discovery on which they would have collaborated. Scientific practices are, today, as little equipped to integrate this new type of rivalry as to struggle against frauds, who put in question the whole of the rules of the game between rival authors.

7. Far from being a defect, this laborious character of the construction of scientific reality is different from the "unilateral" constitutions of "reality" that can be invoked by certain descendants of Kant as well as thinkers who refer to a neurobiological constitution of our "ways" of seeing and anticipating. Here, I am thinking above all of the position of the Chilean biologist Umberto Maturana, which was largely inspired by his works on the perception of frogs. Let me risk a parallel with amphibians. It is easy for us to judge that the "fly" perceived by the frog is only a fiction determined by his neuronal apparatus. By contrast, when the fly is digested, the biologist has to recognize that it is the chemical properties of its constituents, as chemistry has discovered them in its

turn, that are "taken into account," respected and exploited by amphibian metabolism. One could say that the "reality" that scientists seek to bring into existence is closer to that of the digested fly than that of the perceived fly.

8. See studies by Steven Shapin and Simon Schaffer in *Leviathan and the Air-Pump* (Princeton, N.J.: Princeton University Press, 1985).

9. The "void" reveals a private space, the laboratory of "gentleman experimenters," whereas Hobbes means to unify knowledges under the form of an axiomatic capable of constraining any and every one of them to subject themselves, just as he meant to unify civil society under the authority of a sovereign created by contract. Hobbes is thus "Tempier's heir": the axiom, like the sovereign, reveals the power of fiction, but here the fiction, in order to avoid civil war, creates the pseudotranscendence of a fixed point.

10. Bruno Latour, *We Have Never Been Modern*, trans. Catharine Porter (Cambridge: Harvard University Press, 1993), p. 22; translation modified.

11. Ibid., p. 81.

12. In fact, the more powerful the reference, the less resolvable the conflict. Thus, in order to plead for the existence of atoms against Mach's skepticism, Max Planck placed in his camp the "physicist's faith in the unity of the physical world," without which physics would not have been possible, and thus treated Mach as a "false prophet" turning physicists away from their vocation. Likewise, it was when Einstein realized he could not construct an internal critique of quantum mechanics that he suggested condemning it in the name of the hope, which identifies the physicist, of constructing an objective representation of the world, independent of observation. On this subject, see Isabelle Stengers, "Le thème de l'invention en physique," in Isabelle Stengers and Judith Schlanger, eds., *Les Concepts scientifiques* (Paris: La Découverte, 1988; Gallimard, 1991).

13. In his *Galileo Studies*, trans. John Mepham (Atlantic Highlands, N.J.: Humanities Press, 1978), Alexandre Koyré describes this opposition, and shows that Descartes's position toward Galileo was in fact similar to that of Hobbes toward Boyle. In both cases, the philosopher reproaches the scientist for "not thinking," that is, for creating in the laboratory a situation that is unable to give an account of itself in philosophically acceptable terms.

14. This style is already at work when Galileo presents himself as a "midwife," in the Platonic sense, claiming that, in fact, his interlocutors already "know" what he has to teach them (see Koyré, *Galileo Studies*, especially pp. 206–7). However, contrary to Alexandre Koyré, I think that this Platonic argument is not the truth of the

Galilean event (modern physics as a new Platonism), but characterizes its style, in this case the way Galileo distributes, around motion, his adversaries and allies.

15. See Stengers, "Le thème de l'invention en physique." One can maintain that, in its most "technical" aspects, quantum mechanics bears the mark of this disqualification, with regard to what concerns the "leading" [*de pointe*] stakes, by representatives of "phenomenological" physics. On this subject, see Nancy Cartwright, *How the Laws of Physics Lie* (Oxford: Clarendon Press, 1983).

16. This does not contradict the appearance of this other void, the quantum void, which corresponds to completely different experimental apparatuses.

17. See Bruno Latour, *Science in Action: How to Follow Scientists and Engineers through Society* (Cambridge: Harvard University Press, 1988).

18. See Ilya Prigogine, and Isabelle Stengers, *Entre le temps et l'éternité* (Paris: Fayard, 1988).

19. That is, by excluding the pseudoscientific practices that owe their power "to the name of science."

20. This hierarchy is not absolute. In certain cases—for example, when the prestige of a "great program" (the conquest of space, star wars) justifies it—the disciplines accept a more or less egalitarian share of the responsibilities. It is also the case in industrial research, but here the scientist is at risk of losing, in the eyes of his colleagues, what differentiates him from a simple "wage earner."

21. This shows the political dimension of the situation. Quantum chemistry is supposed to be "deducible" from quantum mechanics, whereas the relation is in fact closer to negotiation than to deduction. On this subject, see Bernadette Bensaude-Vincent and Isabelle Stengers, *A History of Chemistry*, trans. Deborah Van Dam (Cambridge: Harvard University Press, 1996).

7. An Available World?

1. For example, it is striking that, in *The Logic of Life: A History of Heredity*, trans. Betty E. Spillmann (New York: Pantheon Books, 1974), François Jacob gives practically no consideration to embryology in the twentieth century. In the perspective instituted by the narrative construction of the triumph of molecular biology, embryology, which was a leading field, has nothing to teach, because it provided nothing that led to the genetic program. Embryology is situated in the future, that is, it must wait for the "upstream movement"—from "bacteria" to the "mouse"—that must be brought about by molecular biology.

2. Jean-Pierre Changeux, *Neuronal Man*, trans. Laurence Garey (New York: Pantheon Books, 1985), p. 125.

3. The fact that the science of engineers had been redefined as "applied science," whose theoretical bases are Galilean mechanics—that is, it was accepted to situate its problems through its "distance from the ideal" that would constitute a world without friction (a world in which the engineer could not work)—passes through a heavy institutional history (conflict between the "inventors" and the Académie des sciences de Paris, in the eighteenth century, creation of the École polytechnique that would become, after the Revolution, the vector of the reorganization of the métier of the engineer in the service of the state).

4. Gilles Deleuze and Félix Guattari, *A Thousand Plateaus: Capitalism and Schizophrenia*, trans. Brian Massumi (Minneapolis: University of Minnesota Press, 1987), for example, p. 159. The judgment of God inspires (p. 161) a warning that brings to mind the Leibnizian principle not to try to reverse established sentiments: "If you free it [the BwO, the body without organs, that is, that which is "divinely" judged in terms of the organism] with too violent an action, if you blow apart the strata without taking precautions, then instead of drawing the plane you will be killed, plunged into a black hole, or even dragged toward catastrophe. Staying stratified—organized, signified, subjected—is not the worst that can happen; the worst that can happen is if you throw the strata into demented or suicidal collapse, which brings them back down on us heavier than ever." To meditate through ironist-sociologists: What will come back down on us, heavier than ever, if they succeed in convincing scientists that their activity is indeed reducible to games of power? To avoid submitting oneself to this judgment and to prudently explore the regimes of coexistence with the network it subsumes, it is recommended to be inspired by the seven "rules of method" and the six "principles" elucidated by Bruno Latour in *Science in Action: How to Follow Scientists and Engineers through Society* (Cambridge: Harvard University Press, 1988), pp. 258–59.

5. The typical example would be the theoretical claim of the "reducibility" of chemistry to the physics of movement and interactions, emitted since the eighteenth century. Each stage of the history in which this claim seems to justify iself signals above all a radical mutation of physics.

6. Judith E. Schlanger, *Penser la bouche pleine* (Paris: Fayard, 1983).

7. In *Lord Bacon* (Paris: Librairie J.-B. Baillière et fils, 1894), Justus von Liebig, one of the inventors of the practice of normal science, erects a veritable inquisition against the notion of "useful" science, which, according to him, was then reigning in England, and links scientific progress, as illustrated by German chemistry, to the refusal to be dispersed in empirical cases judged to be interesting for reasons foreign to science. "An experiment that is not attached in advance to a theory, that is, to an idea, resembles a true investigation as much as the noise of a child's *crécelle* resembles music" (p. 114).

8. For the example of the "reduction" of chemistry to quantum physics, see Bernadette Bensaude-Vincent and Isabelle Stengers, *A History of Chemistry*, trans. Deborah Van Dam (Cambridge: Harvard University Press, 1996).

9. Latour, *Science in Action*, pp. 153–55.

10. Ibid., p. 156.

11. See Deleuze and Guattari, *A Thousand Plateaus*. The rhizome implies the connection between heterogeneous elements: any point can be connected with any other point; it cannot be comprehended in relation with the One, image, project, logic; it can be broken anywhere, and taken up again in accordance with other lines; it cannot be summarized in the name of a genetic principle, but only a cartographic one.

12. It can happen that "error" affects those who would not have to be subjects in it. See Bruno Latour's superb *Aramis, or the Love of Technology*, trans. Catherine Porter (Cambridge: Harvard University Press, 1996), in which the "death of Aramis," a future system of revolutionary transportation, in the end referred to the fact that its "fathers" did not like technology, or were themselves duped by the confusion between sociotechnological innovation and the passage to the existence of an idea, supposed to have in itself the power to realize itself.

13. For the double register of risks, those that one does not have the right to neglect and those that can be delegated to a future where everything will be taken care of "by itself," and for its consequences in the recent history of medicine in the United States, see Diana B. Dutton, *Worse Than the Disease: Pitfalls of Medical Progress* (Cambridge: Cambridge University Press, 1988).

14. See, for example, Isabelle Stengers and Olivier Ralet, *Drogues, le défi hollandais* (Paris: Laboratoire Delagrange, 1991), in which we show that repressive politics toward drugs have, through the selection of adequate experts, hidden the fact that they have no "interest" in addicts who do not define themselves as asking to go off drugs entirely. See also *Drogues et droits de l'homme*, ed. Francis Caballero (Paris: Laboratoire Delagrange/Synthélabo, 1992).

15. For the lucid study of its consequences, whose largely uncontrollable character is now recognized, but accounted for by the "irrationality" of the public, see Michel Tort, *Le Désir froid: Procréation artificielle et crise des repères symboliques* (Paris: La Découverte, 1992).

8. Subject and Object

1. Let me note that *La Nouvelle Alliance*, which was published well before one spoke of "new science," did not plead for such a perspective. The "poetic understanding of Nature" scandalized those who had "forgotten" to read what followed: "in the etymological sense in which the poet is a fabricator"—and who thus confused the idea of the "capacity," of physics, "to respect the nature it makes speak" with the idea of a respect for nature as it is given. See Ilya Prigogine and Isabelle Stengers, *La Nouvelle Alliance: Métamorphose de la science* (Paris: Gallimard, Folio/Essais, 1986), p. 374. [See also the authors' English version of this book, *Order Out of Chaos: Man's New Dialogue With Nature* (New York: Random House, 1984), which differs significantly from the French original.—*Trans.*]

2. For the mythical and anthropological emergence of the object, see Michel Serres, *Statues* (Paris: François Bourin, 1987).

3. The maintenance of the distinction between subject and object implies the maintenance of a distinction between scientific productions and technology. The invention of a technological apparatus cannot, by any approximation, be clarified by the distinction between subject and object, for it has as its matter and as its stake not the identification of what belongs to each of them, but the creation of new modes of dividing them up, which authorizes nothing other than their very possibility. See Bruno Latour, *Aramis, on the Love of Technology*, trans. Catherine Porter (Cambridge: Harvard University Press, 1996).

4. The constructivist thesis according to which all experimentation is "performative," that is, actively creates what occupies the place of the object in it, is "true" from the philosophical viewpoint but disastrous from the practical point of view. If this distinction between viewpoints is neglected, it can wind up weakening all resistance to scientific "pathologies." For example, take the debate that took place in the United States around multiple personalities—are they or are they not produced by the treatment that is supposed to reveal them? The constructivist might be tempted to laugh at the fact that a treatment never "reveals" something that preexists it. But then he is not taking into account the fact that specialists in multiple personalities believe, for their part, that their treatment gives a "really true" truth the power to manifest itself, and that the whole of their practice is authorized by this "really true." Philosophically, the problem of multiple personalities undoubtedly puts in question what we understand by "personality," whether it is an artifact or an intimate truth. (On this subject, see Mikkel Borch-Jacobsen, "Pour introduire à la personnalité multiple," in *Importance de l'hypnose*, ed. Isabelle Stengers [Paris: Synthélabo, 1993].) Practically, this problem must be discussed on the terrain where it is posed, that is, a terrain constituted by the authority of the "really true."

5. See Ed Regis, *Who Got Einstein's Office?* (Reading, Mass.: Addison-Wesley, 1988).

6. Referring, if the case arises, to different disciplines, those that can make simulation an "interdisciplinary" practice.

7. Stephen J. Gould, *Wonderful Life: The Burgess Shale and the Nature of History* (New York: Norton, 1989).

8. Stephen J. Gould, *The Panda's Thumb: More Reflections on Natural History* (New York: Norton, 1980); Stephen J. Gould, *The Flamingo's Smile: Reflections on Natural History* (New York: Norton, 1985); Stephen J. Gould, *Hen's Teeth and Horse's Toes* (New York: Norton, 1983).

9. See the now-classic article by Stephen J. Gould and Richard C. Lewontin, "The Spandrel of San Marco and the Panglossian Paradigm: A Critique of the Adaptationist Program," in *Proceedings of the Royal Society* (London: B205, 1979), pp. 581–98.

10. On this subject, see the the putting in contrast between the sciences of proof and the sciences of indices proposed by Carlo Ginzburg in "Signes traces pistes," *Le Débat* 6 (1980): 2–44.

11. We should not be astonished that paleoanthropology is a privileged terrain for the "demoralization" of history, in this case the history that has "led" to the appearance of Homo sapiens. On this subject, see Roger Lewin, *Bones of Contention* (New York: Simon and Schuster, 1987; London: Penguin, 1991).

12. On this subject, see David M. Raup, *Extinction: Bad Genes or Bad Luck?* (Oxford: Oxford University Press, 1993).

13. The term *actors* is proposed by Bruno Latour to be able to speak in the same manner of humans and of nonhumans, which articulates a controversial situation. Or, here, a simulation on the computer. The definition of the agent [*actant*] is relative to the scence in which he acts; it can be transformed in the course of the narrative, and abandoned to the form of different *actors*.

14. This explains a contrast with regard to which Stephen J. Gould has often expressed his surprise and disappointment: the same interlocutors who would never think of doubting the heliocentric theory or the existence of atoms often consider the set of reconstitutions of the history of living beings coming from paleontology to be irremediably speculative.

15. In which one knows, moreover, the ambiguity of the term. That a team of terrain seeks the means to ameliorate the productivity of a workshop, and almost every experimental means will succeed (transitorily): the interest the workshop's members have for the interest of which they are the object is more determining than the different factors of their "quality of life."

16. See Léon Chertok and Isabelle Stengers, *A Critique of Psychoanalytic Reason: Hypnosis as a Scientific Problem from Lavoisier to Lacan*, trans. Martha Noel Evans (Stanford, Calif.: Stanford University Press, 1992), and Isabelle Stengers, *La Volonté de faire science: À propos de la psychanalyse* (Paris: Synthélabo/Delagrange, 1993).

17. Félix Guattari, *Chaosmose* (Paris: Galilée, 1992).

18. On this subject, see the chapter "Returnings" in Léon Chertok, Isabelle Stengers, and Didier Gille, *Mémoires d'un hérétique* (Paris: La Découverte, 1990).

19. On this subject, see the intrinsic link that Roy Bhaskar proposes to establish between social science and the problematic of emancipation, in *Scientific Realism and Human Emancipation* (London: Verso, 1986).

20. Tobie Nathan, ... *Fier de n'avoir ni pays, ni amis, quelle sottise c'était: Principes d'ethnopsychanalyse* (Paris: La Pensée sauvage, 1993).

21. It is from this latter point of view that one can no doubt speak, by contrast with traditional psychotherapeutic techniques, of a "nonknowledge" proper to psychoanalysis, haunted by the question of the arbitrariness of fiction, and to other contemporary techniques such as Ericksonian hypnosis, which have defended this arbitrariness.

9. Becomings

1. Gilles Deleuze and Félix Guattari, *What Is Philosophy?*, trans. Hugh Tomlinson and Graham Burchell (New York: Columbia University Press, 1994), p. 108.

2. Ibid., p. 110.

3. Ibid.

4. Bruno Latour, *We Have Never Been Modern*, trans. Catherine Porter (Cambridge: Harvard University Press, 1993), p. 144.

5. The idea of a "corporatist" representation obviously has nothing to do with that of the Parliament of Things, because it is inscribed in a static perspective in which stable and well-differentiated groups represent qualified interests in a legitimate manner. The great strength of the parliament of "naked citizens gathered together in the name of the general interest" is that it is able to utilize the corporatist idea as a foil. And the great interest of Latour's hybrids and Guattari's rhizomes, whose common principle is the proliferation and the absence of stable identity, is that they allow one to escape this trap.

6. Deleuze and Guattari, *What Is Philosophy?*, p. 108; translation modified.

7. Ibid.

8. Gilles Deleuze and Félix Guattari, *A Thousand Plateaus: Capitalism and Schizophrenia*, trans. Brian Massumi (Minneapolis: University of Minnesota Press, 1987), p. 369.

9. Ibid., p. 374.

10. On this question, see Bernadette Bensaude-Vincent and Isabelle Stengers, *A History of Chemistry*, trans. Deborah Van Dam (Cambridge: Harvard University Press, 1996).

11. Deleuze and Guattari, *A Thousand Plateaus*, p. 362.

12. In *L'invention des formes* (Paris: Odile Jacob, 1993), Alain Boutot groups these mathematical and physicomathematic innovations (Thom's catastrophes, Prigogine's dissipative structures, Mandelbrot's fractals, Ruelle's chaos, and so on) under the rubric of a "neo-Aristotelianism," opposed in this case to the "dominant techno-science" identified by the author in the manner of Alexandre Koyré and Martin Heidegger. This reading, which immediately associates the scientific style of theoreticians and the philosophical style of their references, nonetheless creates a false symmetry: like Koyré and Heidegger, moreover, Boutot does not take into account the practical dimension (making history) of scientific activity. He sees in these new mathematics "the instrument that is lacking [in the sciences of nature] to apprehend, in its specificity, the moving world of forms, that its complexity renders inaccessible to ordinary quantitative analysis" (p. 314). But he is silent about a "small" difference. The novelty of the mathematical instrument is clear when it concerns forms that hitherto have interested no one: the fall of a leaf, the lizard on a wall, the outline of the coasts of Brittany, and so on; but this "instrument" does not in itself have the power to sustain other ways of working together with regard to "forms" already invested by other practices (cf. Thom's polemical relations with biologists). Moreover, the stagings that oppose the hubris of yesterday's science to the new appprehension, mathematical and pacific, of the world at our scale (which has been carefully depopulated of those stagings, always equally disqualified, that already occupy it), have nothing pacific in themselves, but belong to the ordinary rhetoric of scientific mobilization.

13. From this point of view, the "demobilization" of science can be linked to the question of complexity. On this subject, see Isabelle Stengers, "Complexité: Effet de mode ou problemè?" in *D'une science à l'autre: Des concepts nomades*, ed. Isabelle Stengers (Paris: Seuil, 1987).

14. See Isabelle Stengers and Olivier Ralet, *Drogues, le défi hollandais* (Paris: Laboratoire Delagrange, 1991), and Francis Caballero, ed., *Drogues et droits de l'homme* (Paris: Laboratoire Delagrange/Sythélabo, 1992).

15. See Diana B. Dutton, *Worse Than the Disease: Pitfalls of Medical Progress*, (Cambridge: Cambridge University Press, 1988), pp. 189–92, 319–20.

16. Ibid., p. 320.

17. Félix Guattari, *Les Trois Écologies* (Paris: Galilée, 1989), p. 46. I have purposely selected here the citation that allows Luc Ferry, in *The New Ecological Order*, trans. Carol Volk (Chicago: University of Chicago Press, 1995), p. 113, to accuse Guattari of undermining the "values of the *res publica*."

18. In *We Have Never Been Modern*, Bruno Latour announces the possibility of thinking the one without forgetting the other from the concept of "transcendence that lacks a contrary": "The world of meaning and the world of being are one and the same world, that of translation, substitution, delegation, passing" (p. 129). Gilbert Simondon's work creates an analogous perspective from the concept of transduction, on the condition that the task of the "philosopher-technologist" he fondly appeals to is not (as Gilbert Hotois believes in his useful presentation, *Simondon et la philosophie de la "culture technique"* [Brussels: De Boeck-Université, 1993]) a simple affair of "thought" identifying the dissociations that are due to the sole insufficiency of traditional culture, but the "transductive transduction" of an effective, aesthetic, ethical, and political mutation, which refers to the challenge of the "Parliament of Things." As for myself, I will explain this perspective one day in terms derived from the philosophy of Alfred North Whitehead.

19. Latour, *We Have Never Been Modern*, p. 97.

20. Or rather, according to Latour, the art of a resistance that cannot pride itself on having no transcendence, since transcendence is without a contrary.

Index

abstraction, 78, 86
Académie des sciences, 175
acceleration, 83
actor, 11, 143; in Latour, 177. *See also* author
adaptation, 139–41. *See also* evolution
addiction, 160. *See also* drugs
administrative sciences, 27
aesthetics, 92, 148
affectivity, 92
Africa, 98
ahistoricity, 39
AIDS, 160
air pump, 100–103. *See also* apparatus; Boyle, Robert
Alembert, Jean le Rond d'. *See* d'Alembert
alliance: in primate societies, 63
Alliez, Eric, 79
Althusser, Louis, 25, 27
Alvarez, Luis, 143
Alvarez, Walter, 143
ambulant processes, 155–56
America: discovery of, 97–98
anarchism, 37
animal, 22; versus human, 62
anomaly, 6
Antarctic, 153
anthropology, 5, 8
antibiotics, 4, 68
apes, 146

Apocalypse, 143
apparatus, 90, 132; as "black box," 103, 120; Boyle's
    air pump, 100–103; creates "facts" and "readers,"
    96; experimental, 84; Galileo's inclined plane,
    84–87, 90, 100; lacking in the field sciences,
    140
Aquinas, Thomas: *Summa Theologica*, 93
Aramis, 176
Arendt, Hannah, 61, 95, 172
argumentation, 41; and fiction, 79; in Galileo, 72
Aristotle, 77, 178; and *phronesis*, 94; on politics, 61
art criticism, 29
art history, 41
artifact, 84, 92, 120, 138, 147; and fact, 51
astonishment: as motif, 72
astrology, 12, 35
asymmetry, principle of, 65
atomic bomb, 4
atoms, 156
Augustine, 69, 163
author: and authority, 93; two definitions of, 89. *See
    also* actor
authority, 126; and author, 93; creation of, 95; and
    history, 93–94
autonomy, 13, 105; of community, 8; norm of, 3; of
    scientific community, 5
availability, 112
Avogadro's number, 115

baboons, 63, 146, 173
Bachelard, Gaston, 25–27; on epistemological breaks, 25;
  on opinion, 26
Bacon, Francis, 175
bacteria, 108, 111, 171
Barberini, Cardinal Maffeo, 77
Bateson, Gregory, 45
becoming, 93, 148, 151–67; minoritarian, 170; of
  scientists, 92; social, 129; and truth, 15, 167; woman,
  162
Beethoven, Ludwig van, 39
before and after, 118
belief, 60; loss of, 152; mobilizing, 114
Benjamin, Walter, 171
Bensaïd, Daniel, 171
Bensaude-Vincent, Bernadette: *Etudes sur Hélène
  Metzger*, 41
Bernal, John D.: *The Social Function of Science*, 7
Bernalism, 7
Bernard, Jean, 129
Bhaskar, Roy, 172, 177
bias, 11
Bible, 73
big bang, 98
bioethics, 12
biology, 171; evolutionary, 139; molecular, 11, 112
black box, 103, 120. *See also* apparatus
Bloor, David, 172
body: in medicine, 24–25; body without organs, 175
Bohr, Niels, 107
Borch-Jacobson, Mikkel, 176
Bordeu, Dr., 17
Boudon, Raymond, *The Art of Self-Persuasion*, 170
bourgeois science, 7
Boutot, Alain, 178
Boyle, Robert, 100, 102–4, 154, 174. *See also* air pump
break: versus continuity, 52; versus demarcation, 25–28;
  epistemological, 25
bricolage, 105
Brownian particle, 115
Bukharin, Nikolay, 7
Butler, Samuel, 15; *Erewhon*, 78, 135

calculability, 12–13, 124; and science, 10
Cambridge Experimentation Review Board, 161
Candide, 140
Carnap, Rudolf, 26
Carnot, Nicolas Sadi, 39
case: nature of, 115, 136
catastrophe theory, 178
cause versus reason, 45
Chalmers, Alan, 27, 31
Changeux, Jean-Pierre, 112, 153
chaos, 178
charletanism, 23–24
chemistry, 11, 25, 31, 42; reducibility to physics, 175–76

China, 98
Christianity, 78–79
Church, 72
CIA, 144
citizen, 60
city: and politics, 72
civilization, 4
claimants, legitimate, 61
cognitive structures, 23–24, 149
Cohen, Daniel, 129, 153
cold fusion, 174
cold war, 144
colleagues: role of, 90, 154
colonization, 152
Columbus, Christopher, 42, 97
communication: in Habermas, 100; resistance to, 155;
  sciences of, 27
community, scientific: autonomy of, 49, 105; in Kuhn, 5;
  as norm, 5, 7
competition: in scientific research, 11
computers, 112; computer simulation, 136
concept: in Deleuze and Guattari, 154; and philosophy,
  151
confirmation, 30; in Popper, 29
Congress on the History of Science and Technology
  (1931), 7
conquest: principle of, 154
consensus, 8
construction, social, 119
constructivism, 97
consumers: power of, 125
contingency: aesthetic of, 145; in science, 71–72
controversy, 92; role of, 90
convention, 4, 30, 33; human, 13; in Popper, 90
Copernicus, Nicolaus, 21–22, 32
cosmology, 31, 98; Aristotelian, 77; Kantian critique of,
  100–101
counterpower, 118
creationism, 132
Crick, James, 4
critiques of science: empiricist versus radical, 21; feminist,
  11; technoscientific, 10–11. *See also* technoscience
culture: human versus animal, 62; versus science, 10;
  Western, 148

d'Alembert, Jean le Rond, 112
Darwin, Charles 139, 148, 158; Darwinism, 43, 132,
  141–42, 167; neo-Darwinism, 139. *See also* evolution
data: production of, 22
Davis, Ray, 171
definitions: abstract, 91; as fiction, 76
Deleuze, Gilles, 15, 127; on concepts, 154–55; on
  foundation versus ground, 68; on rationality, 71;
  rhizomes, 124, 126, 177; and shame, 151; *A Thousand
  Plateaus*, 113, 155, 170. 175–176; *What is Philosophy?*,
  71. *See also* Guattari, Félix; judgment, of God

demarcationist tradition, 21–37, 50, 82; versus break, 25–28; criteria of, 29. *See also* Popper, Karl
demobilization of science, 178
democracy, 61, 158, 160
demythification of science, 14
Descartes, René, 21–22, 102, 174
determinism, 48
Diderot, Denis, 17, 112
difference, 6, 13, 16, 45, 58, 60, 67, 158; assessing, 48; as criterion of demarcation, 29; empirical, 60
DNA, 107, 116
doctors versus charletans, 23–24
Drake, Stilman, 83, 173
drugs, 176. *See also* addiction
Duhem, Pierre, 78, 173; *The Aim and Structure of Physical Theory*, 29
Dutch drug addicts, 160

Eco, Umberto: *Foucault's Pendulum*, 143; *The Name of the Rose*, 143
École polytechnique, 175
economics, 27
efficacy, 124
Eichmann, Adolf, 23
Einstein, Albert, 9, 14, 28, 29, 30, 35, 44, 47, 174; on Mach and positivism, 28
Eldredge, Niles, 140
embryology, 111
empirical traits, 60; empiricist critique of science, 11, 21
Endocrinology (journal), 120, 122–23
engagement, 30, 92, 132
England, 7; sociology in, 3
Enlightenment, 10, 153
epistemological break. *See* break, epistemological
epistemology, 21–22, 42, 43, 60, 91; constructivist, 97; error of, 124; Feyerabend's farewell to, 37; French, 25; normative, 74
error, 27, 44, 176, 124–25; versus mistake, 125
essence versus fiction, 80
ethics, 11, 131, 147–48
ethics committees, 11, 17
ethnology, 58
Europe, 9, 15; and America, 97–98; four centuries of science, 9
evaluation: of research criteria, 7
event, 50, 51, 67–69, 104, 116, 145, 163; as abstraction, 86; and authorship, 89; concept of, 67; double constraint of, 164; and fiction, 80; and history, 96; and mediation, 99; versus right, 156
evolution, 43, 139; adaptation, 139–41; natural selection, 43, 140; survival of the fittest, 140–41. *See also* Darwin, Charles
existence: bringing into, 92, 126, 131, 147, 144–45; production of, 146, 162
experiment, 22, 32, 49; experimental sciences, 47; Galileo's inclined plane, 83–87, 90, 100

expertise: production of, 159–62
extinctions, 143

facts, 6, 27, 30–31, 78; and artifact, 51; creation of, 86; impregnated by theory, 33, 50; invention of, 50; matters of, 102; observable, 49; production of, 51; and reasoning, 32; and theory, 49
faith, 79
falsification, 30, 32, 89; in Popper, 28, 89
feminism, 21, 162, 132–33; critique of scientific rationality, 10–11
Feyerabend, Paul, 42, 46, 48, 57, 60, 78, 91, 106, 108, 115; *Against Method*, 35–37
fiction, 115, 148; assessing the difference, 131; and computer simulation, 136; as definition, 76; and event, 80; mathematical, 135–39; power of, 79, 95, 145, 75–80; as taboo, 153
field sciences: nature of, 144
Fleischman, Martin, 174
force: relations of, 9, 12, 37, 50, 58; as nonscientific factors, 8
form: and matter, 156
foundation: versus ground, 68, 118. *See also* Deleuze, Gilles; ground
fractals, 157, 178
France, 91; sociology in, 3
Frank, Philip, 26
Freud, Sigmund, 25, 147–48, 158. *See also* psychoanalysis
Freudenthal, Gad, 41

Gadamer, Hans-Georg, 41
Galileo, Galilei, 21–22, 35, 42, 72, 80–81, 90, 94, 101–7, 112, 128, 133, 135–37, 139, 157–58, 163, 173–74; concept of speed, 83–87; condemnation, 77; *Dialogue*, 75–80, 85, 141; *Discourse concerning Two New Sciences*, 75–77; the inclined plane, 83–87, 90, 100
Généthon program, 129
genetics, 111; genetic code, 115
Gillispie, Charles: and history of the sciences, 7
Ginzburg, Carlo, 177
God, 22, 67; and freedom, 79; and grace, 69; and Stephen Hawking, 81; omnipotence, 77–78
Gould, Stephen J., 139–42, 177; *Wonderful Life*, 141–42
grace, theory of, 69
Greece, 46, 61, 163; and origins of philosophy, 71; and politics, 61; and rivalry, 71; and science, 78
greenhouse effect, 144, 159
ground, 14, 86; versus foundation, 68, 118. *See also* Deleuze, Gilles; foundation
Guattari, Félix, 127; *Chaosmosis*, 170; new aesthetic paradigm, 148; on the production of subjectivity, 162; on rationality, 71; and shame, 151; *A Thousand Plateaus*, 113; *What is Philosophy?*, 71. *See also* Deleuze, Gilles

Habermas, Jürgen, 100
Hacking, Ian, 172; on "intervening," 49

Harding, Sandra, 133–34, 149; *The Science Question in Feminism*, 21
Harvard University, 161
Hawking, Stephen, 81–82
Hayes, Dan, 161
healing practices, 23
Hegel, Georg, 9, 164
Heidegger, Martin, 61, 178, 163–65; on science, 101
Heisenberg, Werner, 4
heliocentrism, 32, 65
heterogenesis, 162
hierarchy, 126; in science, 108
history, 11; and authority, 93–94; construction of, 9; demoralizing, 142–45; and event, 96; in evolution, 140; force of, 39–53; hermeneutical style, 41; internalist, 9; internal versus external, 7–8, 39, 171; making, 146, 89–108
history of science, 32; internal versus external, 7–8, 39, 171; singularity of, 39
Hobbes, Thomas, 154, 174; *Leviathan*, 154; on the vacuum, 100
holistic science, 132
human sciences: as pre-paradigmatic, 52
Hume, David, 21–22
humor, 53, 164, 167; definition of, 66; and irony, 18, 57–69; as political project, 66–67
hybrids, 99, 177
hylomorphism, 155
hypnosis, 25; and psychoanalysis, 170

ideal: distance from, 13, 91; of "true" science, 8–9
ideology, 4, 27, 79, 104; in Althusser, 27
idols, 30
immanence: and humor, 66. *See also* transcendence
inclined plane, 81, 90, 100–101, 163; in Galileo, 83–87. *See also* apparatus; Galileo, Galilei
incommensurability: between paradigms, 49
industry, 9, 128
information technology, 10, 112, 136, 161
innovation, 9, 117–18, 128
interest, 13, 92, 95, 116, 126; experimental, 91; impure, 9; nonscientific, 13; practical, 156; of scientists versus philosophers, 8; social and economic, 7
intersubjectivity, 100; and existence, 101
intervening, 137; in Hacking, 49
invention, 6, 80, 115, 128, 131, 145, 158; experimental, 89; of problems, 156; scientific, 9
irony, 53; and humor, 18, 57–69; as political project, 66–67; and relativism, 68; stable versus dynamic, 66
irrational, 158; construction of history, 9
irrational numbers, 46, 155
irreduction: Latour's principle of, 16, 61, 91

Jacob, François, 175
journalism, 122
judge, 16, 29, 152, 164; power to, 60, 113

judgment, 143; of God, 113–14, 154, 175; of priority, 8; of scientists, 158; of value, 8

Kant, Immanuel, 21–22, 30, 73, 133, 100–101; *Critique of Practical Reason*, 30; *Critique of Pure Reason*, 30, 173; and the new science, 73; and phenomena, 82
Kepler, Johannes, 73
knowledge: development of, 8; ideal of, 61; production of, 146; scientific, 8, 11; theory of, 22
Koyré, Alexandre, 94, 106, 173–74, 178; and history of the sciences, 7
Kuhn, Thomas, 4–6, 33, 48, 52–53, 57, 68, 72, 78, 93, 106, 117, 170, 172; concept of the paradigm, 48–53; and Hegel, 9; *Structure of Scientific Revolutions*, 8. *See also* normal science; paradigm

laboratory, 9, 22, 85, 90, 93–94, 104, 112, 121, 131, 154, 156, 166; philosophers and historians excluded from, 90; role in science, 92
Lacan, Jacques, 163
Lakatos, Imre, 28, 31, 32, 33, 37, 48, 78, 107; on Popper, 28–29
Lamarck, Jean Baptiste, 140
language: and observation, 29
Laplace, Pierre Simone de, 112
Latour, Bruno, 16, 66, 87, 91, 99–100, 103, 119, 121, 124, 154, 159, 171, 173; *Parliament of Things*, 152–55, 157–59, 162; *We Have Never Been Modern*, 17, 65, 100, 124, 178
laughter, 18, 53, 114, 131
Lavoisier, Antoine Laurent, 170; *Method of Chemical Nomenclature*, 106
Law, Judaic, 67
laws: of motion, 74; of nature, 163
legitimacy: of power and knowledge, 62
Leibniz, Gottfried, 15, 102; and God, 78; his "shameful declaration," 15; the Leibnizian constraint, 15–18, 58, 97; and monadology, 69
Lespinasse, Mlle. de, 17
Leviathan, 154. *See also* Hobbes
Lévi-Strauss, Claude, 4
Liebig, Justus von, 96; on Bacon, 175
life, 26, 31; artificial, 138
living beings, 44
Locke, John, 21–22
logic, 27, 30, 43, 45, 46; operative, 12; in Popper, 29; and positivism, 26
logos: in Greece, 61
London, 7
Lorentz, Konrad, 30

Mach, Ernst, 28, 174
magnitude, problem of, 135
Mandelbrot, Benoit, 157, 178
Mannoni, Octave, 170
market, free, 7, 9; in Polanyi, 9
Marx, Karl, 148, 158; Marxism, 7, 25, 27, 29–30, 44

Masterman, Margaret, 172
mathematics, 11, 18, 46, 86, 102, 178, 135–39; fractal, 157; nomadic, 157; Popper on, 155; theorematic, 157
matter: and form, 155–56
matters of fact, 102
Maturana, Umberto, 172, 174
meaning: absence of, 12; and Popper's second world, 45; and signification, 45
measure, 163, 166; concept of, 163; man as "the measure of all things," 162–63; and measurement, 27; mode of, 164
mechanics: Galilean, 175; rational, 102
mediators, 99–103
medicine, 11, 23–24, 113–14, 121
memory, cultural, 117
Mesmer, Franz Anton, 170. *See also* hypnosis
metaphysics, 27, 28; positivistic rejection of, 26
method, 8; scientific, 112
Metzger, Hélène, 41, 51; *La Chimie*, 41
microbes, 42, 145; and Pasteur, 96–97
Middle Ages, 77, 79, 116, 135
Milgram, Stanley, 22–23
minority: in Deleuze and Guattari, 170
mobilization, 104, 126, 144, 154, 114–19; defined as a form of organization, 10; paradigmatic versus technological, 117–18; and rhetoric, 156
model: defined, 135; versus theory, 135
modern: category of, 65
Monod, Jacques, 107, 113; *Chance and Necessity*, 111
moral: definition of, 143; law, 95; morality, 27
motion: in Galileo, 80, 82; three types of, 84
movement: mathematical definition of, 90
multiplicity, 137; of interpretations, 67
mutants, 43

narration, 151–52, 167; creation of, 142–43; in evolution, 141
Nathan, Tobie, 148
National Academy of Sciences, 120
National Institutes of Health, 161
natural selection, 43, 140
nature: made to speak, 81, 89; and mathematics, 107
Needham, Joseph, 7
networks: and power, 126; politics of, 124–29
neurobiology, 174
new: category of, 6, 43, 49
Newton, Isaac, 39, 73, 94, 112, 133
Nobel prize, 120, 171
nomad science, 155–56
noncommensurability: thesis of, 4
nonscience, 21–37; as opinion, 27
normalization, 13
normal science, 4, 33, 117. *See also* Kuhn, Thomas
norms, 29; methdological, 8; scientific, 74; search for, 80, 91

nuclear war, 144, 146; nuclear winter, 143
numbers, irrational, 46

object: and subject, 131–49, 176
objective knowledge: in Feyerabend, 35
objectivity, 105–6; category of, 16; construction of, 35; in Feyerabend, 36; imperative of, 40; as mask, 9; as "myth," 60; norm of, 3–4; scientific, 15, 23
observation, 78; and language, 29; production of, 50
open society, 28
opinion, 61, 71, 133, 163; and ideology, 27; and politics, 163; versus science, 26, 132
Oregio, Cardinal, 76–77
other, 11

paganism, 79, 127
pandorine, 119–20
Pangloss, Dr., 140
paradigm, 4–6, 8, 14, 105; aesthetic, 148; as an anomaly, 6; incommensurability between, 49; in Kuhn, 48–53; paradigmatic crisis, 6; as tacit, 5; versus vision of the world, 111. *See also Kuhn*, Thomas
parapsychology, 12, 92. *See also* psychology
Parliament of Things, 152–55, 157–59, 162. *See also* Latour, Bruno
passion, 30; two meanings of, 166
Pasteur, Louis, 96; and microbes, 97
paternalism, 164
patron: role of, 119–24, 153
Pelagianism: on grace, 69
pendulum, 73, 86
performative, 176
Perrin, Jean, 42, 112, 115
personality, 176
phenomena, 115; creation of, 51, 99; saving, 137
phenomenology: in Polayni, 6
philosophy, 124, 151; origins of, 71
philosophy of science, 14; status of, 8
phronesis, 94
physics, 11, 25, 31, 86; and chemistry, 175–76; fundamental versus phenomenological, 104; high-energy, 173; phenomenological versus theoretical, 102
Piaget, Jean, 4
Pinch, Trevor, 172; *Confronting Nature*, 171
placebo effect, 24–25, 147
Plato, 61, 72, 94–95, 102, 163, 174–75
poiesis, 62, 94
Poincaré, Henri, 30
Polanyi, Michael, 6–7, 9; creation of Society for Freedom in Science, 7; "The Republic of Science," 7
political science, 58–59
politics, 163; as "everybody's business," 59; Greek definitions, 61, 66–67; and opinion, 163; of reason, 64; and scientific reason, 15; and sociology, 60; two types of, 106; versus sociology, 58
politologists, 62

Pons, Stanley, 174
Popper, Karl, 27–28, 35, 43–48, 51, 57, 78, 162, 170; and falsification, 28, 89; *The Logic of Scientific Discovery*, 28–29; on the open society, 28; on the situation, 44; the three Poppers in Lakatos, 28–30; three worlds, 42–48, 63, 155, 172; *The Unrealized Quest*, 43. *See also* demarcationist tradition
positivism, 26–27, 44; in Feyerabend, 36; in Popper, 29
Pouchet, Félix-Archimède, 96
power, 12, 61; of consumers, 125; as an effect of the event, 104; in Latour, 124; and network, 126; repressive, 79; social, 9
praxis, 62, 163; and Aristotle, 94
Prigogine, Ilya, 170, 176, 178
primatology, 62–63; 146
principle. *See* asymmetry; conquest; irreduction; symmetry
private versus public, 60
problematization, 61
problems, 44, 47–48, 57, 128; contemporary, 60; invention of, 156; political character, 157; posing of, 91. *See also* questions
procedure: objective scientific, 22; truth, 92
process: contingent, 71–72
production: of data, 22; of existence, 162; of expertise, 158–62; of facts, 51; of knowledge versus existence, 146–47; of observation, 50; of research, 6; scientific, 7–8; of subjectivity, 162; technological, 10; of theory, 108; of truth, 92
profitability, 13
progress, 8, 31, 81, 151, 152; category of, 16; image of, 152; scientific, 4, 6, 22
proletariat science, 7
propositions: scientific, 27
Protagoras, 162
psychism, 148
psychoanalysis, 27, 29–30, 44, 121–22, 147–48. *See also* Freud, Sigmund; hypnosis.
psychology, 11, 46; mob, 5. *See also* parapsychology
public: as nonscientific, 23; versus private, 60
putting to the test, 134
puzzle solver, 50

quantum chemistry, 175
quantum mechanics, 113–114; and Einstein, 174
questions, 52; how and why, 81–82; political, 103; posing, 6. *See also* problems

rational: how scientists define, 128
rationality, 4, 124; category of, 16; feminist critique of, 10; Freudian, 25; normative, 35; operative, 127; in Popper, 29; scientific, 10, 14; of scientists, 8. *See also* reason
rats, 146; use of, 22
realism, 4, 31, 50, 76
reality: category of, 16

reason: communicative, 100; cunning of, 9; definition of, 34; in Greece, 61; labor of, 35; in medicine, 23–25; new use of, 80–83; operative, 22, 112; politics of, 64, 160; versus cause, 45. *See also* rationality
reduction: power of, 116
refutation, 60; in Popper, 29
relativism, 76
relativity, 28
religion, 15. *See also* God
representation, 163, 135–36; abstract scientific, 86; mathematical versus experimental, 135; scientific, 86–87
research, 169; choice of priorities, 5; as production, 6; scientific, 11
research programs, 33, 49–50
resistance, 11, 151; art of, 166
retaliation: argument of, 3–4, 10
revolution, 53, 90; paradigmatic, 5
rhetoric, 14, 115, 156
Rhine, Joseph B., 92
rhizome, 124, 126, 177
right: divine, 80; rights of citizenship, 60
risk, 111–12, 134, 158
rivalry, 9, 80; in Greece, 71; in scientific research, 11
robotics, 138
Rome, 73, 82, 141
Rouch, Jean, 58
Roudinesco, Elisabeth, 170
Ruelle, David, 178

Sagredo, 75–77, 82–83, 86, 91
Saint John of the Cross, 121, 122
Salviati, 75–77, 86
Saussure, Ferdinand, 4
Schlanger, Judith, 41, 117
Schlick, Moritz, 26
Schrödinger cat, 114
science: administrative, 27; as ahistorical, 39; applied, 175; autonomy of, 105; bourgeois, 7; condition of possibility of, 9; as contingent process, 71–72; definition of, 25, 26, 29; destructive, 10; and event, 71–87; experimental, 47; field, 47, 140–41, 144; hierarchy within, 104; history of, 7, 65; holistic, 132; human, 5; identified as "theoretico-experimental," 132; impure, 3, 13; leading versus applied, 113; mathematical, 47; "in the name of," 21–25, 28, 69, 99, 111, 126, 158, 164; new, 132; and nonscience, 21–37, 44, 64, 104; versus nonscience, 26; normal, 4, 33, 117; versus opinion, 26; political invention of, 64–67; proletariat, 7; pure versus applied, 3, 14, 118; rational, 8, 10; royal versus nomad, 155–56; singularity of, 40; soft, 23; true, 13
scientificity, 124
scientific type: in Popper, 28
scientist: competent, 7; and controversy, 9; heroic, 43; as rational, 5
Scripture, 73

selection, natural, 43, 140. *See also* evolution
sentiments, 15. *See also* Leibnizian constraint
Serres, Michel, 176
shame, 151
signification: and meaning, 45–46
Simondon, Gilbert, 178
Simplicio, 75–77
simulation, 136–38. *See also* computers
singularities, 156
singularity, 24, 31, 64; definition of, 134; of science, 95; of
    the sciences, 58, 131–35
situation, 46, 48; experimental, 49; in Popper, 44
skepticism, 6; in Middle Ages, 79; relativist, 75
social history, 8
society: definition of, 59–60; and exclusion, 63–64;
    problematized, 61
Society for Freedom in Science, 7
sociologists: "doing their job," 10, 58–59
sociology, 3–7, 42, 83; and the model of the positive
    sciences, 59; and politics, 60; versus politics, 58;
    relativist, 10, 12; of science, 10, 59; of the sciences, 9
Sophists, 61, 134; new, 136; return to, 162–67
space: in Einstein, 28
speed, 82, 154; three concepts of, in Galileo, 83–87
Sri Lanka, 172
state: and science, 9
statement: versus theory, 112
strong program, 172
Strum, Shirley, 63, 163, 173
subject: as free, 133; and object, 131–49; versus object,
    176
subjectivation, 162
Sunflower, Professor, 154
survival of the fittest, 140–41. *See also* evolution
symmetry: principle of, 8, 12
systems theory, 5

tacit: character of paradigms, 5; knowledge, 6
Talmud, 67
Taminiaux, Jacques, 95, 174
tautology: in demarcationist tradition, 30–31
technē, 94, 163
technology, 10, 152; instrumental, 22
technoscience, 10, 12, 17, 26, 112, 118, 165; critiques of,
    108
telescope: and Galileo, 73
Tempier, Étienne, 77, 79, 127, 147, 174
territory, 118–19; concept of, 170
Testart, Jacques, 129
theatricalization of the Earth, 143

theory: definition of, 107; versus experimental statement,
    112; and facts, 33, 49; hyperbolic, 170; versus model,
    135; and observation, 29; power of, 6; production of,
    108
thermodynamics, 39
third party, 156
third world, 157, 160
Thom, René, 157, 178
three little pigs, story of, 159–60
three worlds, 42–48, 63, 155, 172. *See also* Popper, Karl
time: in Einstein, 28
torture of animals, 22–23
transcendence, 16, 66, 95, 162
transcendental philosophy, 22. *See also* Kant, Immanuel
transferral: in Popper, 43
tribunal: of the laboratory, 133
True, 95; in Greece, 61
truth, 15, 18; and becoming, 167; category of, 16; defini-
    tion of, 158; in medicine, 23–25; as "myth," 60; neces-
    sary truths versus truths of fact, 79; negative, 89–93,
    103; new type introduced by Galileo, 73–75, 90; in
    philosophy, 163; in Popper, 30; and procedure, 92;
    production of, 92; as unveiling and denunciation, 17

unconscious, 147, 153
Urban VIII, Pope, 77
utilitarianism, 62

vacuum, 94, 102; vacuum pump, 103. *See also* air pump
Varela, Francisco, 172
variable: independent, 85
variation, 145
Vellucci, Alfred, 161
Venus, 32
Vernant, Jean-Pierre, 61
Vienna Circle, 26–27
vision of the world, 111. *See also* paradigm
vivisection, 22
void, 174; quantum, 175

Western science: in Feyerabend, 37
Whitehead, Alfred North, 178; and Leibniz, 169
windmill, 90, 174
witness, 147; faithful, 111, 115, 131, 135, 167
Wolfram, Steve, 136
women, 133. *See also* feminism
Woolgar, Steven, 66
worlds, three. *See* Popper, Karl

Yung Lo, Chinese emperor, 98

Isabelle Stengers is associate professor of philosophy at the Free University of Brussels and a Distinguished Member of the National Committee of Logic and the History and Philosophy of Sciences in Belgium. She is the author of numerous books, including *Order Out of Chaos* (with Ilya Prigogine) and *Power and Invention: Situating Science* (Minnesota, 1997). She received the Grand Prix de Philosophie from the Académie Française in 1993.

Daniel W. Smith is a postdoctoral fellow in philosophy at the University of New South Wales in Sydney, Australia. He has translated Gilles Deleuze's *Essays Critical and Clinical* (with Michael A. Greco; Minnesota, 1997) and Pierre Klossowski's *Nietzsche and the Vicious Circle*.